Modeling Crop Rotations and Co-Products in Agricultural Life Cycle Assessments

Gerhard Brankatschk

Modeling Crop Rotations and Co-Products in Agricultural Life Cycle Assessments

Gerhard Brankatschk
Fachgebiet Sustainable Engineering
Technische Universität Berlin
Berlin, Germany

Zugl.: Berlin, Technische Universität, Diss., 2017

Additional material to this book can be downloaded from http://extras.springer.com.

ISBN 978-3-658-23587-1   ISBN 978-3-658-23588-8   (eBook)
https://doi.org/10.1007/978-3-658-23588-8

Library of Congress Control Number: 2018955424

Springer Vieweg
© Springer Fachmedien Wiesbaden GmbH, part of Springer Nature 2019

This Springer Vieweg imprint is published by the registered company Springer Fachmedien Wiesbaden
GmbH part of Springer Nature
The registered company address is: Abraham-Lincoln-Str. 46, 65189 Wiesbaden, Germany

Contents

List of Figures

List of Tables

Summary

Life cycle assessments (LCA) are applied to evaluate environmental interventions associated with products and services whilst considering entire supply chains and full life cycle. Agricultural LCAs are used to assess production processes of agricultural raw materials and products derived thereof i.e. food, feed, fibre and fuel. Whenever the environmental performance is influenced by differences in agricultural practices, the LCA model should be able to reflect these agricultural details, in order to express the actual impact of the products, produced in different agricultural systems and to assist improving the environmental performances of these agricultural systems. Several methodological challenges in LCA of agricultural systems do exist, e.g. soil quality, biodiversity, land use, spatial and temporal differentiation. The focus of this work is contributing to better represent multifunctionality (co-products) and temporal system boundaries (crop rotations) in agricultural LCAs. Research targets are defined aiming at representing multifunctionality, avoiding non- and double counting, integrating animal and vegetable production, developing an biophysical agriculture-specific denominator, including long-term and phytosanitary effects amongst crops, comparing different or modified crop rotations, remaining the product focus whilst considering crop rotations, visualizing the numerical consequences, providing all data required for the use of new methods and ensuring compatibility to ISO standards for LCA. In order to solve the lack of an agriculture-specific co-product allocation approach for LCA and the lack of adequately representing crop rotations in LCA, two new life cycle inventory (LCI) approaches were developed.

Firstly, a new LCI-approach for co-product allocation is elaborated, described and compared to established methods. This Cereal Unit allocation approach is based on animal-nutritional value and proves to be suitable for a large range of agricultural products and co-products, including vegetable and animal products. Conversion factors for more than 200 products are provided. Compared to established allocation approaches, the shares for allocating environmental burdens between wheat grain and wheat straw are as follows: mass allocation 56% (grains)/44% (straw), energy allocation

55%/45%, economic allocation 77%/23% and Cereal Unit allocation 75%/25%. Applying the Cereal Unit allocation approach to the question of allocating between milk, calf and diary cow leads to comparable results (86.6%, 6.6% and 6.8%, respectively) with other biological allocation approaches.

Secondly, the crop rotation approach, a new LCI-approach is elaborated for including entire crop rotations – and thus the interactions between crops, grown in temporal succession on the same field – into agricultural LCAs. The ability of performing product-oriented assessments is retained. Comparison of LCI results for wheat grains grown in a rotation of sugar beet – barley – wheat – rapeseed – wheat reveals Nitrogen (N) demand of 25.6 kg N / t versus 30.9 kg N / t in current LCA practice and without considering the crop rotation. For rapeseeds, the difference is 32.0 kg N / t in crop rotation versus 54.4 kg N / t in current LCA practice.

Both of the LCI-methods are tested in a case study. Product carbon footprints (PCF) are calculated for wheat bread, cow milk, rapeseed biodiesel and wheat straw-bioethanol; N demand was used as variable. Consideration of crop rotations for wheat bread, cow milk and rapeseed biodiesel leads to lower PCFs (-11%, -22% and -16%, respectively). Even larger variations (-34% to +95%) were reported from other research groups, applying the presented LCI method. Consideration of straw as an agricultural co-product leads to higher PCF for wheat straw-bioethanol (+80%), compared to calculation practice, prescribed in EU-legislation, in which crop residues are considered as waste – ostensibly being available without any environmental burdens. Relevance of crop residues to soil quality and soil fertility is explained.

In future, the need for assessing differences between agricultural management options will increase. Crop rotations are one of the oldest agricultural management tools. Their relevance towards climate-smart agriculture will increase. Farmers will need assistance in assessing the environmental performances of different management options. Politicians' and consumers' interest in environmentally sound bioeconomy products (food, feed, fibre, fuel) will increase – and therefore the need for product-based LCA.

Presented LCI methods allow modeling agronomic interrelations of agricultural systems within life cycle inventories. They help assessing temporal, spatial and multifunctional complexity of agricultural systems and help improving the reliability of life cycle based sustainability assessments of agriculture. The new LCI methods do not affect other LCA stages, e.g. impact assessment, or the functional unit, and they are compatible to ISO-standards for LCA. Research groups from more than ten countries have already started using the methods.

Particularly challenging for future agriculture is being simultaneously productive and climate-smart. Presented methodologies incorporate the performance principle, which allows LCA optimizing agricultural systems towards improved quotient of environmental burden per production. This work provides solutions for methodological limitations of agricultural LCAs and thus contributes to the challenging process towards sustainable agriculture.

Zusammenfassung

Ökobilanzen (engl. Life Cycle Assessment) dienen zur Bewertung von Umweltwirkungen, die bei Herstellung und Anwendung von Produkten und Dienstleistungen entstehen. Hierbei werden Lebensweg und Lieferkette in ihrer Gesamtheit einbezogen. Landwirtschaftliche Ökobilanzen werden zur Bewertung der Erzeugung von Agrarrohstoffen und daraus hergestellter Produkte durchgeführt – beispielsweise für Lebensmittel, Futtermittel, Nachwachsende Rohstoffe sowie Bioenergie. Sofern die Umweltwirkungen landwirtschaftlicher Erzeugnisse von der Art und Weise der jeweiligen Landbewirtschaftung abhängen, sollten die landwirtschaftlichen Praktiken in der Ökobilanz auch angemessen abgebildet sein. Nur hierdurch können Unterschiede zwischen verschiedenen Produktionssystemen sichtbar gemacht und sachgerechte Empfehlungen zu deren Verbesserung abgegeben werden. Hierbei existieren mehrere methodische Herausforderungen – etwa die Berücksichtigung von Bodenqualität, Biodiversität, Landnutzung, sowie die räumliche und zeitliche Auflösung. Schwerpunkt dieser Arbeit ist es, einen methodischen Beitrag zur Verbesserung der ökobilanziellen Berücksichtigung von Multifunktionalität (Nebenprodukte) und zeitlicher Systemgrenzen (Fruchtfolgen) landwirtschaftlicher Systeme zu leisten. Folgende Forschungsziele wurden hierfür definiert: Multifunktionalität berücksichtigen, Nichtzählung und Doppeltzählung von Umweltwirkungen vermeiden, tierische und pflanzliche Produktion gleichzeitig berücksichtigen, biophysikalischen landwirtschaftsspezifischen gemeinsamen Nenner entwickeln, Langzeiteffekte und phytosanitäre Effekte zwischen Feldfrüchten berücksichtigen, Vergleich unterschiedlicher oder modifizierter Fruchtfolgen ermöglichen unter gleichzeitiger Beibehaltung des Produktbezugs (einzelnes landwirtschaftliches Erzeugnis), Einfluss der neuen Methoden quantifizieren, notwendige Daten zur Anwendung der neuen Methoden bereitstellen und Kompatibilität zu den ISO-Normen für Ökobilanzierung wahren. Um die Lücken des fehlenden landwirtschaftsspezifischen Nebenprodukt-Allokationsverfahrens zu schließen und die mangelnde Abbildbarkeit von Fruchtfolgen in Ökobilanzen zu ermöglichen, wurden zwei neue Methoden entwickelt, die innerhalb der Sachbilanz (engl. Life Cycle Inventory) anzuwenden sind.

Erstens wurde ein Nebenprodukt-Allokationsverfahren für die Sachbilanz entwickelt, beschrieben und mit bestehenden Allokationsverfahren verglichen. Dieses Getreideeinheiten-Allokationsverfahren basiert auf dem tierischen Futterwert und hat sich für die Darstellung der großen Bandbreite landwirtschaftlicher Produkte und Nebenprodukte – inklusive pflanzlicher und tierischer Erzeugnisse – als geeignet erwiesen. Hierfür werden mehr als 200 Umrechnungsfaktoren bereitgestellt. Der Vergleich mit etablierten Allokationsverfahren zur Aufteilung der Umweltwirkungen zwischen Weizenkorn und Weizenstroh hat folgende Anteile hervorgebracht: Masseallokation 56% (Korn)/44% (Stroh), Energetische Allokation 55%/45%, Ökonomische Allokation 77%/23% und Getreideeinheitenallokation 75%/25%. Die Anwendung der Getreideeinheitenallokation auf die Frage der Aufteilung der Umweltwirkungen zwischen Milch, Kalb und Milchkuh ergibt vergleichbare Ergebnisse (86,6%, 6,6% und 6,8%) zu jenen anderer biologischer Allokationsverfahren.

Zweitens wurde mit dem Fruchtfolgeansatz ein neues Sachbilanz-Verfahren zur Integration der gesamten Fruchtfolge in Ökobilanzen entwickelt. Hierdurch werden auch Wechselwirkungen zwischen den in zeitlicher Aufeinanderfolge auf demselben Feld angebauten Früchten berücksichtigt. Der Vergleich der Sachbilanz-Ergebnisse für Weizenkörner, die in einer Fruchtfolge aus Zuckerrübe – Gerste – Weizen – Raps – Weizen angebaut wurden ergibt einen Stickstoff (N) Bedarf von 25,6 kg N / t im Gegensatz zu 30,9 kg N / t in bisheriger Praxis ohne Berücksichtigung der Fruchtfolge. Für Rapssaaten betragen die Werte 32,0 kg N / t in der Fruchtfolge im Gegensatz zu 54,4 kg N / t in gegenwärtiger Ökobilanz-Praxis.

Diese beiden Sachbilanz-Methoden wurden in einer Fallstudie untersucht. CO_2-Fußabdrücke (engl. Product Carbon Footprint) wurden für Weizenbrot, Kuhmilch, Raps-Biodiesel und weizenstrohbasiertes Bioethanol berechnet und der Stickstoffbedarf als Variable verwendet. Die Berücksichtigung der Fruchtfolge bei Brot, Milch und Biodiesel führte zu geringeren CO_2-Fußabdrücken (-11%, -22% und -16%). Eine andere Forschergruppe, von der vorgestellter

Fruchtfolge-Ansatz bereits eingesetzt wurde, haben noch größere Abweichungen (-34 bis +95%) festgestellt. Am Beispiel weizenstrohbasierten Bioethanols führte die Einbeziehung von Stroh als landwirtschaftliches Nebenprodukt zu größeren CO_2-Fußabdrücken (+80%) im Vergleich zur gemäß EU-Gesetzgebung vorgeschriebenen Berechnungsweise, in der landwirtschaftliche Reststoffe pauschal als Abfall betrachtet werden und vermeintlich ohne Umweltwirkungen zur Verfügung stehen. Die Bedeutung landwirtschaftlicher Reststoffe für die Erhaltung der Bodenqualität und Bodenfruchtbarkeit wird erläutert.

In Zukunft wird der Bedarf steigen, ökobilanzielle Bewertungen unterschiedlicher Landbewirtschaftungsoptionen vorzunehmen. Fruchtfolgen gehören zu den ältesten landwirtschaftlichen Praktiken. Es ist abzusehen, dass ihre Rolle für klimafreundliche Landwirtschaft immer wichtiger wird. Landwirte werden Unterstützung benötigen, um die Umweltwirkungen verschiedener Bewirtschaftungspraktiken zu bewerten. Das Interesse von Politikern und Verbrauchern in umweltfreundliche Produkte der Bioökonomie (Lebensmittel, Futtermittel, Nachwachsende Rohstoffe und Bioenergie) wird steigen – und mit ihm auch der Bedarf produktbezogener Ökobilanzen.

Die vorgestellten Sachbilanz-Methoden ermöglichen, die Zusammenhänge landwirtschaftlicher Systeme in Ökobilanzen zu integrieren. Sie werden dabei behilflich sein, die zeitliche, räumliche und multifunktionelle Komplexität von Agrarsystemen zu bewerten. Gleichzeitig werden sie die Verlässlichkeit lebenszyklusbasierter Nachhaltigkeitsbewertungen der Landwirtschaft verbessern. Die neuen Sachbilanz-Methoden beeinflussen weder eine andere Phase der Ökobilanz, wie etwa die Wirkungsabschätzung, noch die Funktionelle Einheit. Darüber hinaus sind sie kompatibel zu den ISO-Normen für Ökobilanzierung. Wissenschaftlergruppen aus mehr als zehn Ländern haben bereits begonnen, die vorgestellten Methoden zu verwenden.

Eine ganz besondere Herausforderung für die Landwirtschaft der Zukunft ist es, gleichzeitig produktiv und klimaschonend zu sein. Die vorgestellten Methoden enthalten das Leistungsprinzip, welches der Ökobilanz ermöglicht, landwirtschaftliche Systeme hin zu einem verbesserten Quotienten aus Umweltwirkung pro Produktion zu optimieren. Diese Arbeit liefert Lösungen für methodische Schwachstellen der Ökobilanzierung landwirtschaftlicher Systeme und leistet daher ein Beitrag auf dem anspruchsvollen Weg hin zur Nachhaltigen Landwirtschaft.

1 Introduction

Sustainable development is a multidimensional process and global challenge for humankind. One of the key challenges during global development process is maintaining environment in conditions that allow prosperous life for future human generations. As the state of the environment – air, water, soil, organisms – brings decisive prerequisites to the quality of human life, it is rational to preserve environmental conditions. All human activities lead to environmental interventions. Therefore, one of the key questions for human development is identifying and promoting environmentally sound options in order to maintain basic living conditions.

Food is an essential living condition to all humans on earth. Any provision of food is causing a certain environmental intervention, whereas different life cycle stages, i.e. agriculture, transport and food preparation, contribute to quality and quantity of the intervention. A major contributor to the environmental impact of food is the agricultural production. The agricultural stage of a specific agricultural raw material (e.g. wheat grains) is characterized by geographical variations (e.g. climatic conditions, soil properties, water availability) and by a large number of different agricultural practices (e.g. fertilization strategies, soil management, farming systems). This leads to varying environmental impacts for one agricultural raw material, produced in different processes. A closer look at agricultural processes is required to understand and to improve their environmental interventions. Life cycle assessment (LCA) is an internationally standardized and established tool for evaluating environmental interventions of processes and to derive recommendations for reducing environmental burdens [1].

Applying LCA to agricultural systems allows reducing the environmental burdens for all products derived from agriculture. Besides food, as well feed, fibre and fuel are produced from agricultural raw materials. Agricultural LCA are therefore relevant for many products, based on biomass. Several governments grant political support for products and services that provide good environmental performances on a life cycle basis. In this context, LCA or approaches based on LCA, such as product carbon footprints

© Springer Fachmedien Wiesbaden GmbH, part of Springer Nature 2019
G. Brankatschk, *Modeling Crop Rotations and Co-Products in Agricultural Life Cycle Assessments*, https://doi.org/10.1007/978-3-658-23588-8_1

(PCF), are already established for assessing environmental interventions of agricultural products and products derived thereof. Examples are biofuel legislations in several regions in the world that require minimum greenhouse gas (GHG) savings from biofuels compared to fossil fuels.

Accuracy of agricultural LCA is far from being perfect. The LCA community has identified a lot of potential to improve the accuracy and robustness of agricultural LCA results. The next section describes key challenges for agricultural LCAs.

1.1   Challenges in Life Cycle Assessments for Agricultural Systems

Despite large efforts in the past, a large number of scientific publications, case studies and the dedicated conference series LCA Food [2-8], which already took place ten times between 1996 and 2016 [9], there are still several challenges and open questions for agricultural LCAs. These challenges can be categorized in data availability or quality issues and methodological issues.

LCA has been traditionally used for production, packaging, fabrication or manufacturing processes. Here, the inputs, the outputs, the system boundaries and the functional units can often be defined in relatively clear terms. With regard to the inputs (fertilizers, fuel, lubricants, crop protection agents and seedling material), the situation of agricultural production is comparable to fabrication processes. But fundamental challenges emerge when it comes to the spatial dimension, temporal dimension and multi-functionality of agricultural systems. Methodology of LCA is not yet well enough elaborated to tackle all special characteristics of agricultural systems – leading to situation in which the LCA models have limitations for reflecting agricultural systems. Models always need to simplify the reality to some extend in order to be manageable, which is valid for LCA as well. But with

regard to agricultural systems and their assessments towards future challenges of agriculture, there is a need for further fine adjustment within LCA
methodology. Representing different agricultural management options will be
necessary to assess differences amongst them and to derive robust recommendations for farmers towards more sustainable farming practices. In 2017,
Notarnicola et al. elaborated an overview of key methodological challenges
for LCA of agri-food systems [10]. Based on that review article, challenges
are described in the next paragraphs.

The **spatial dimension** of agricultural production sites is not clearly
defined and a "distinction between technosphere and ecosphere in relation to
modeling of environmental impacts" is missing [10]. "Soil fertility, soil
structure, soil hydrology balance [and] soil biodiversity" are temporarily not
included, even though affected by agricultural activity [10].

This leads to the question, which **impact categories**, affected by agricultural systems, are considered within LCA. „Decreased soil quality and
fertility, increased erosion, reduced ecosystem services due to intensification,
biodiversity loss ..." are not captured by LCA and should be „... addressed in
order to further advance the currently available approaches and
methods."[10] "A possible solution to overcome this flaw is to include
agricultural soil in the ecosphere or to include the evaluation of these impacts under the land use impact category." [10] Further impact categories of
relevance are "water use, biodiversity, toxicity, particular matter, where as
the robustness of these impact assessment methods should be improved. As
long as these aspects are not considered in impact categories, shifts between
impact categories remain undetected. For example, a farm with good greenhouse gas performance would be positively assessed in current LCA practice,
whilst not detecting potential impacts to soil quality and soil fertility that
are not considered at all. Two further relevant modeling challenges are
"inconsistencies between emission inventory modelling and impact assessment of pesticides" and the "assessment of land use change associated with
off-farm inputs to agricultural production systems." [10]

"A consensus is still missing on a globally applicable model for calculating soil and water **emissions**", i.e. nutrient leaching, soil erosion and run-

off. These models for the inventory level should consider local soil conditions, such as pH, clay content and slope. [10]

Among and even within **individual agricultural enterprises**, a high intrinsic variability does exist. "Different **management practices**, **soil types** and **climates**, **seasonality**, the life cycle of **perennial crops**, and **distances** (and related transport modes) between locations ... [are causing] ... considerable variability in inventory data ..." [10]. Inventory databases do not cover all of those variabilities. Because „databases are usually created using data representing specific sites at specific times ... [they] are **not interchangeable** with each other and need to be used with caution by LCA practitioners ... to allow a fair and meaningful comparison of food production systems, a high level of geographical specificity is needed for agri-food systems. " [10] Furthermore, differences within and between "**management practices** [of] ... 'organic', 'biodynamic', 'integrated', 'heated greenhouse' production" [10] do exist and should be considered.

Agricultural operations may produce **different qualities** of products, e.g. ice wine and wine. How to deal with different qualities of agricultural products? The definition of a **functional unit** is not easy – is it a nutritional function or is it a cultural function?

The **multifunctionality** is another important aspect. Agriculture has **multiple outputs (co-products)**, whilst just one of the outputs might of interest for a product-based LCA. The LCA practitioner needs to allocate the environmental burden amongst specific co-products that are typically used in very different sectors and the allocation has potential to largely influence the LCA result. Established allocation approaches using lower heating value (energy allocation), market price (economic allocation) and product weight (mass allocation) can be unable to reflect the diverse functions of all co-products at the same time. Therefore, in agricultural LCAs, different allocation approaches are applied by LCA-practitioners. Several authors did suggest to develop a sector-specific allocation approach [11], that is based on biophysical criteria [12]. This challenge is called co-product allocation problem.

The **temporal dimension** of agricultural systems is another crucial aspect. Typically, the temporal system boundary in agricultural LCAs is defined as one year – representing one vegetation period. In contrast to this life cycle modeling procedure, farmers' agricultural perennial management decisions are characterized by a longer time horizon than one year. **Crop rotations and cropping systems** have been developed centuries ago [13], were key for increase of yields and were therefore relevant for the development of humankind. Today, crop rotations continue to play an essential role in good agricultural management practice. Amongst management tools, crop rotations play a central role in climate-smart agriculture and resource efficient agriculture. Crop rotations lead to improved phytosanitary conditions, which allow reducing the use of pesticides. Crop rotations enhance nutrient efficiencies, which allow reducing the use of fertilizers and is limiting the loss of fertilizers by leaching. Crop rotations help reducing climatic and economic risks to farmers, because a broader spectrum of crops is less vulnerable to bad weather conditions and market price fluctuations. Accordingly, crop rotations influence the need of inputs and hence, are relevant for the environmental performance of the assessed agricultural production system.

The relevance of crop rotations is underlined by the *Sustainability Assessment of Food and Agriculture Systems* (SAFA)-indicators of the *Food and Agriculture Organization of the United Nations* (FAO); SAFA indicators refer to the need of:

- "Extended crop rotations" as GHG mitigation practice,
- "Diverse crop rotation, including the introduction of fodder and cover crops" as soil improvement practice, for maintaining the soil chemical quality and for land conservation,
- "Longer crop rotations, including nitrogen fixing species" as ecosystem enhancing practices,
- "Crop rotation and the production of several crops and species simultaneously" as product diversification and
- "Keeping pests at a minimum level through crop rotations" as tool against hazardous pesticides. [14]

Although the qualitative relevance of crop rotations is unquestioned, there are no quantitative tools available to assess and compare these effects among different crop rotation approaches for the production of one specific product. Farming strategies, such as crop rotation planning, intermediate cropping, multiple cropping, multi-annual fertilization and improving phytosanitary conditions are not considered in current LCA practice. Hereby, life cycle assessment fails assessing differences in long-term management decisions and thus ignores mid- and long-term effects of agricultural practices. This methodological weakness limits LCAs capability for deriving robust recommendations towards long-term agricultural management practices.

Within this work, not all of the previously mentioned challenges of agricultural life cycle assessments are tackled. Rather a focus is given to the aspects of multifunctionality, co-product-allocation, temporal dimension and crop rotations. For these challenges, the next section describes research questions and research targets.

1.2 Research Questions and Research Targets of This Work

Within the previous section, several challenges in LCAs for agricultural systems have been identified. Based on the challenges of multifunctionality, co-product-allocation, temporal dimension and crop rotations, the following **research questions** are derived:

- How can the multifunctionality and the multiple outputs of agricultural systems being taken into account in LCA?
- How can the unintended double-counting or non-accounting of environmental interventions being avoided, which are caused by different co-product allocation approaches?
- How can vegetable and animal production being captured in one allocation approach, e.g. for farm LCAs?

- How to integrate crop rotation effects in agricultural LCAs, e.g. nutrient shift, yield impact and phytosanitary conditions?
- How to evaluate modifications in crop rotation systems, e.g. introduction of legumes, and how to identify environmentally sound crop rotations?
- How to assess improved environmental performances of agricultural products, derived from different crop rotations?

This work aims at contributing to the solutions of these questions. In order to support methodological improvement of agricultural LCAs, the following **research targets** are defined:

1. Consider multifunctionality of agricultural processes
Agricultural production is characterized by multifunctionality and multiple outputs. An approach should be developed to take multifunctionality and multi-output of agricultural raw material production systems and primary processing of agricultural products into account.

2. Avoid unintended double- or non-accounting of environmental burdens
Unintended double-counting or non-accounting of environmental interventions occur, whenever two independent LCAs are performed for one and the same process (e.g. harvesting, grain milling or oilseed crushing) and each of the LCA uses different allocation approaches. This might not be the case for ISO-conform LCA practice, but is being observed when using LCA methodology within political frameworks with fixed allocation rules (e.g. European Renewable Energy Directive or Product Environmental Footprinting). An allocation approach should be developed, which can be universally applied to agricultural production and hereby reduces incentives for choosing different allocation approaches – this will help avoiding unintended double-counting or non-accounting of environmental interventions.

3. Consider vegetable and animal products in one allocation approach
Frequently, farms produce vegetable and animal products at the same time.

For LCAs of entire farms, LCA methodology should be able to fairly attributing environmental burdens between vegetable and animal products (field crops and livestock products). An allocation approach should be made available which allows including both, vegetable and animal products at the same time.

4. Common agriculture-specific denominator based on biophysical mechanisms
Using mass, energy content and economic value as basis for allocation lead to ignoring product's quality (mass) or potentials for use (energy) or affected by subsidies (economic). An agriculture-specific allocation approach should be developed that is based on biophysical relationship – it should be able representing a large share of all agricultural products and easy to use in agricultural LCA.

5. Mid- and long-term effects of agricultural management strategies and improved phytosanitary conditions
The discrepancy should be overcome between system boundaries in current LCAs (typically, one year or one vegetation period) and the multi-annual relevance of agricultural management decisions that unfold their effects over several years. For instance in long-term fertilization free-rider phenomena occur because crops benefit from fertilization, performed year(s) ago, but not accounted in current assessment – due to the system boundary, limited to one year. Furthermore, targeted improvements of phytosanitary conditions are currently not considered in LCA – therefore, strategies for reduced use of pesticides are represented to a limited extend. A new methodological framework should be developed which goes beyond one vegetation period and recognizes mid- and long-term effects of agricultural management decisions into LCA.

6. Comparing environmental performances of different crop rotations
Farmers will need robust recommendations for their crop rotation planning towards environmentally sound decisions. Agricultural LCAs should become able to compare environmental performances of different crop rotations and

modified crop rotations. A methodological framework should be developed to depict differences among crop rotations in LCA.

7. Assess effects of integrating legumes, fallow and multiple cropping
A methodological framework should be developed that allows assessing the environmental impact of integrating legumes, enhanced crop diversity, new crops, multiple crops (either two or more crops or agroforestry), intermediate crops, cover crops or fallow in agricultural LCAs. This is necessary to verify the contribution of these management options towards improvements of environmental performance. The effects might be relevant both, on the level of the entire rotation and for each of the crop rotation elements individually.

8. Product-focus when assessing different crop rotations
Frequently, LCAs are performed to express environmental performance of a single product. The agricultural raw material, used to produce this single product, might be produced in different production systems, i.e. different crop rotations, leading to different environmental performances. A methodological framework should be developed which allows reflecting that products might be originate from different crop rotations (monoculture, short rotation, long rotation, et cetera) in agricultural LCAs. This will allow comparing environmental performances for products from different crop rotations and therefore, setting incentives for environmentally improved cropping systems.

9. Numerical consequences of applying new approaches
Numerical implications of new methods are not obvious from the method itself. Therefore, there is a need to apply new methodological approaches to case systems and to compare the results with those, obtained using current modeling practices.

10. Ensure compatibility to standardized LCA methodology (ISO 14040 series) and availability of data for immediate use and further development
Approaches should be ready-to-use and suitable for further improvement. All conversion factors, background data and steps for further methodological improvement should be made available. Approaches should not create a completely new framework of environmental assessments. They rather shall contribute to advancement of existing LCA and being compatible to the International Standard ISO 14040 series.

These ten research targets are tackled within the next chapter. Aiming at contribution to both, methodological improvement and applicability of new methods, the results chapter contains a method-oriented part and an application-oriented part.

2 Results

This cumulative thesis essentially consists of research articles[*] and related supplementary documents. Therefore, the results section of this work consists of five sub-sections, containing research articles, supplementary data or supplementary material. New methods are being elaborated and explained within sections 2.1, 2.2 and 2.3 Applicability of the new methods is tested within sections 2.4 and 2.5.

The sections in detail:
– Section   2.1   Application   of   the   Cereal   Unit in A New Allocation Procedure contains the publication:

Brankatschk, G., & Finkbeiner, M. (2014). Application of the Cereal Unit in a new allocation procedure for agricultural life cycle assessments. Journal   of   Cleaner   Production,   73,   72-79.   doi: http://dx.doi.org/10.1016/j.jclepro.2014.02.005

---

[*] List of publications:

Brankatschk, G., & Finkbeiner, M. (2014). Application of the Cereal Unit in a new allocation procedure for agricultural life cycle assessments. Journal of Cleaner Production, 73, 72-79. doi: http://dx.doi.org/10.1016/j.jclepro.2014.02.005 [15]   G.   Brankatschk   and   M.   Finkbeiner, "Application of the Cereal Unit in a new allocation procedure for agricultural life cycle assessments," *Journal of Cleaner Production,* vol. 73, no. 0, pp. 72-79, 2014.

Brankatschk, G., & Finkbeiner, M. (2015). Modeling crop rotation in agricultural LCAs — Challenges   and   potential   solutions.   Agricultural   Systems,   138,   66-76.   doi: http://dx.doi.org/10.1016/j.agsy.2015.05.008 [16]   G.   Brankatschk   and   M.   Finkbeiner, "Modeling crop rotation in agricultural LCAs — Challenges and potential solutions," *Agricultural Systems,* vol. 138, no. 0, pp. 66-76, 9// 2015.

Brankatschk, G. & Finkbeiner, M. (2017). Crop rotations and crop residues are relevant parameters for agricultural carbon footprints. Agronomy for Sustainable Development, vol. 37, no. 6, p. 58; doi: http://dx.doi.org/10.1007/s13593-017-0464-4 [17]   G. Brankatschk and M. Finkbeiner, "Crop rotations and crop residues are relevant parameters for agricultural carbon footprints," *Agronomy for Sustainable Development,* vol. 37, no. 6, p. 58, 2017/10/30 2017.

© Springer Fachmedien Wiesbaden GmbH, part of Springer Nature 2019
G. Brankatschk, *Modeling Crop Rotations and Co-Products in Agricultural Life Cycle Assessments*, https://doi.org/10.1007/978-3-658-23588-8_2

–   Section 2.2 Supplementary Data – Application of the Cereal Unit contains the Supplementary Material, belonging to the publication:

Brankatschk, G., & Finkbeiner, M. (2014). Application of the Cereal Unit in a new allocation procedure for agricultural life cycle assessments. Journal   of   Cleaner   Production,   73,   72-79.   doi: http://dx.doi.org/10.1016/j.jclepro.2014.02.005

–   Section 2.3 Modeling Crop Rotation in Agricultural LCAs contains the publication:

Brankatschk, G., & Finkbeiner, M. (2015). Modeling crop rotation in agricultural LCAs — Challenges and potential solutions. Agricultural Systems, 138, 66-76. doi: http://dx.doi.org/10.1016/j.agsy.2015.05.008

–   Section 2.4 Crop Rotations and Crop Residues are Relevant for Carbon Footprints contains the publication:

Brankatschk, G. & Finkbeiner, M. (2017). Crop rotations and crop residues are relevant parameters for agricultural carbon footprints. Agronomy   for   Sustainable   Development,   vol.   37,   no.   6,   p.   58;   doi: http://dx.doi.org/10.1007/s13593-017-0464-4

–   Section   2.5   Supplementary   Material   –   Crop   Rotations and Crop Residues contains Supplementary material, which provides background information and the life cycle inventory that was elaborated for the publication:

Brankatschk, G. & Finkbeiner, M. (2017). Crop rotations and crop residues are relevant parameters for agricultural carbon footprints. Agronomy   for   Sustainable   Development,   vol.   37,   no.   6,   p.   58;   doi: http://dx.doi.org/10.1007/s13593-017-0464-4

Each subsection is introduced by a short description of the content and an assignment, which of the previously defined research questions is addressed. An overview of which target is addressed in which section is presented in **Table 1**.

Table 1 Outline: Assignment of research targets to results chapters

Research targets		Chapters				
		2.1	2.2	2.3	2.4	2.5
#	Short description	Publication Cereal Unit allocation	Supplementary data belonging to 2.1	Publication crop rotation approach	Publication case studies agricultural PCFs	Supplementary material LCI of 2.4
1	Multifunctionality	√				
2	Avoid non- and double counting	√			√	
3	Animal and vegetable products	√	√	√		
4	Biophysical agriculture-specific denominator	√	√			
5	Long-term and phytosanitary effects			√	√	
6	Different crop rotations			√		
7	Modified rotations (introducing new rotation elements)			√		
8	Product focus when assessing rotations			√	√	
9	Numerical consequences	√		√	√	√
10	Data availability and ISO compatibility	√	√	√	√	√

2.1　Application of the Cereal Unit
　　in A New Allocation Procedure
　　for Agricultural Life Cycle Assessments

This section contains the publication:

Brankatschk, G., & Finkbeiner, M. (2014). Application of the Cereal Unit in a new allocation procedure for agricultural life cycle assessments. Journal of Cleaner Production, 73, 72-79. doi: http://dx.doi.org/10.1016/j.jclepro.2014.02.005

The paper introduces the problems of multifunctionality, co-product allocation and unintended double- or non-accounting of environmental burden as source of uncertainty in agricultural LCAs (contributing to research targets #1 and #2). The Cereal Unit is described as common denominator for agricultural systems (#3,4). As a contribution to the LCA methodology, the new agriculture-specific Cereal Unit allocation approach is elaborated and introduced. Step-by-step instructions are provided (#1,2,3,4) and illustrative life cycle inventory examples are given. Applying and comparing the Cereal Unit allocation approach to the multi-output system 'milk-calf-cow' revealed comparable results to other biological allocation approaches (#4,9). Furthermore, comparison of the new allocation approach to established allocation approaches, i.e. economic, energy and mass allocation, is provided (#9). The new Cereal Unit allocation approach is intended for use within the life cycle inventory (LCI), as part of the internationally standardized LCA method (#10).

Abstract

The results of life cycle assessments (LCA) can be significantly affected by the choice of allocation procedure because different allocation approaches lead to a wide range of results. Agricultural systems are particularly sensitive due to their co-products being used in various sectors and accounted for at several allocation steps. If the allocation procedures for different products from the same agricultural system are not aligned to one another, methodological inconsistencies might occur. Parts of the environmental burden might be either unaccounted or doubly accounted for. As a consequence, the overall environmental burden of the agricultural system is not properly assessed.

The Cereal Unit (CU) has been used as a common denominator in German agricultural statistics for decades and is mainly based on the nutritional value for livestock. Products and co-products not intended for livestock feeds are also covered. More than 200 CU conversion factors are provided for vegetable and animal products and co-products occurring during their processing. To calculate the CU, the specifically aggregated metabolizable energy content is calculated for each feed material and normalized using barley as a reference (1 kg barley 1/4 12.56 MJ specifically aggregated metabolizable energy). The CU approach can be applied to other regions without prohibitive efforts.

In this paper, we derive an allocation approach that is based on the CU as an existing unit for agricultural products. The new CU allocation approach is tested and compared to established allocation approaches for wheat, barley, soybean, rapeseed, sugar beet and sunflower. The CU allocation generates results between the approaches of mass, energy and economic allocations. For instance, the allocation shares between wheat grain and wheat straw are as follows: mass allocation 56% (grains)/44% (straw), energy allocation 55%/45%, economic allocation 77%/23% and CU allocation 75%/25%.

We demonstrate that the CU is an appropriate unit for the description of agricultural products and can serve as the basis for an agriculture-specific allocation approach in LCA. CU allocation may help to address agricultural allocation problems and might lead to more robust LCA results for products and services originating from raw agricultural materials. We recommend further testing and future application of this new allocation approach.

2.1.1   Introduction

In recent years, the need for an accurate quantification of the environmental impacts of products and services has grown rapidly, as evidenced by enhanced information requirements in the supply chain of products [18, 19] and increased public awareness and communication of environmental footprints, such as carbon footprints [20], in addition to full LCAs. One risk associated with this development is the fact that most consumers and policy makers are not fully aware of the uncertainty of LCA results related to methodological choices. Driven by the carbon footprint discussion, LCAs for agriculture have gained increasing interest. There are methodological particularities and challenges in agricultural LCAs, reflected, for instance, in specific conference series, such as the International Conference series on Life Cycle Assessment in the Agri-Food Sector – LCA Food [5].

Agricultural LCAs contain products and co-products that cover a broad range, from cereal straw to fattening calf. Allocation procedures are used to attribute environmental burdens between the products. ISO 14044 describes a hierarchy of allocation approaches, which are preferably based on scientific aspects (e.g., mass allocation or lower heating value allocation) rather than economic relationships (e.g., market price allocation) [21]. For the sake of credible results, LCA practitioners try to treat all products and co-products as fairly and as adequately as possible, but existing allocation approaches focus on the specific functionalities of individual co-products and do not reflect in all cases the purposes of all co-products at the same time. Agricultural LCAs are particularly affected because allocation steps often take place several times and errors introduced by each allocation step propagate. An adequate allocation approach is crucial for the credibility of the LCA results of this sector. This aim could be achieved by finding an allocation solution that reliably represents the common functions provided by the wide range of agricultural products. Lundie et al. [11] recommend establishing sector-specific allocation procedures based on physico-chemical relationships. Pelletier and Tyedmers [12] suggest biophysical criteria as a basis for co-product allocations. They argue that allocation approaches should be causality driven and should express the motivation for a certain activity. Furthermore, biophysical approaches are more flexible and can be better adapted to the motivation and causalities of the processes [12]. Within this paper, we focus on co- product allocation within LCAs for agricultural products and products derived from raw agricultural materials.

2.1.1.1 Need for an Allocation Procedure in Attributional LCA Modeling for Agricultural Products

Various approaches have been developed for allocating environmental burdens among multiple inputs and outputs. Wellknown examples include mass allocation, energy allocation and economic allocation. Alternatives that avoid allocation include system expansion, subdivision or substitution [21, 22]. The standards ISO 14040 and ISO 14044 give guidance on how to address allocation situations, but they offer a hierarchy of choices rather than a particular method [1, 23]. The ISO hierarchy leaves room for different allocation procedures, whereas other LCA standards, such as PAS 2050 and BP X 30-323-0, give clear recommendations for the use of economic allocation or physical allocation, respectively [24, 25]. Such contradictory hierarchies within the LCA-based standards for attributional modeling complicate the allocation choices.

Additional strategies for dealing with co-products exist within the consequential LCA modeling approach that need to be treated separately from attributional LCA modeling. Ekvall and Weidema [26] demonstrate the use of system expansion to handle the co-products of renewable materials leading to the approach of consequential LCA [26-28]. The allocation problem can be avoided in consequential LCAs by applying system expansion, but system expansion and avoided burden approaches cause additional uncertainties [11]. Consequential LCAs do not fall within the scope of this paper.

For attributional LCAs, the allocation problem, both in general and in particular for agricultural products, is still a topic of concern and debate; several studies provide more detailed analyses of this topic [11, 23, 29-32]. The approach presented in this paper is intended for use in attributional LCA modeling for agricultural products.

The next two sections describe situations where the application of allocation procedures may introduce uncertainty to the results or even lead to situations where environmental burden might be double counted or even ignored.

2.1.1.2   Different Allocation Methods as Source of Uncertainty

The different approaches for co-product allocation are one of the major reasons for the uncertainty in LCA results caused by methodological choices [33-36]. To address the allocation problem, various strategies have been developed, including system expansion, system reduction, allocation based on physical causality, mass allocation and economic allocation, but none of them are completely satisfactory [37].

Agricultural LCAs are particularly sensitive because allocations are carried out several times in succession. For example, during harvesting, the wheat plant split into grains and straw. During the milling process, wheat grains are split into flour, bran and middlings. Therefore, the inaccuracies introduced by each allocation step propagate and potentially amplify. For the calculation of agriculture-based chains, Chiaramonti and Recchia [38] describe "dramatic variations [of the results] (up to approximately 300% or more if different approaches toward co-product allocation are considered). This will happen even in case a very simple and small biofuel chain [.] is considered". Another example pertaining to allocation choice is provided by Cavalett and Ortega [39], who performed a case study for soybean biodiesel; they conclude that the allocation choice is a "very significant calculation step" that "strongly affects" the results. Using different allocation methods, Luo et al. [40] compare the environmental effects of gasoline and bioethanol. The outcome is fundamentally affected by the choice of the allocation method. The results were even inverted by changing the allocation method from economic to mass or energy allocation. Against the backdrop of having different allocation approaches as sources of uncertainty, Lundie et al. [11] recommend developing physico-chemical and sector-specific allocation procedures. Pelletier and Tyedmers [12] advocate for biophysical criteria that simultaneously reflect physical properties and social functions.

2.1.1.3 Unintended Ignoring or Double Counting of Environmental Burden due to Different Allocation Methods

Another consequence of the allocation problem is the phenomenon of ignoring or double counting environmental burden. This methodological artifact occurs if the allocation approaches of two (or more) LCAs that contain co-products grown in the same agricultural system are not aligned to each other. The sum of the sub-systems' burdens is not equal to the total environmental burden of their underlying agricultural production process. If both sub-systems are considered in one LCA study, this should not happen because ISO 14040 and 14044 requires use of the same allocation approach [21, 22].

However, if two co-products from the same production process are considered within different studies, e.g., they are used in different sectors, the LCAs are calculated independently. The individual LCA practitioners are not restricted in their decision of allocation approach (e.g., wheat flour and wheat bran in the food and feed sectors; vegetable oil and oilmeal in the bioenergy and feed sectors). If those practitioners are using different LCA standards (ISO 14040 and 14044, PAS 2050 or French BP X 30-323-0) or sector-specific LCA guidelines (European Renewable Energy Directive, International Dairy LCA Guide), it is reasonable to expect different allocation approaches to be used, even if the co-products originate from the same production process. As a result, the phenomenon of ignoring or double counting environmental burden is likely to occur.

For example, dairy production and biodiesel production use rapeseed meal and rapeseed oil. Both are most likely derived from the same rape seeds because, e.g., in Germany, the use of rapeseed meal for animal feeding and the use of rapeseed oil as a raw material for biodiesel production are quantitatively the most important uses [41-43]. Typically, Life Cycle Assessments of dairy and biodiesel are performed separately. Relevant guidance documents recommend the use of different allocation methods for the same rapeseed processing step. The International Dairy Federation (IDF) advises in its LCA guideline "to use economic allocation for co-product in feed production" [44], whereas the European renewable energy directive requires that "greenhouse gas emission shall be divided between fuel or its intermediate product and the co-products in proportion to their energy content" [45]. Therefore, it is very likely that different allocation methods are being used for meal and oil, even though they may originate from the same rape seeds.

To illustrate the phenomena of ignoring and double counting environmental burden in quantitative terms, two allocation choice scenarios for determining the environmental burdens of rapeseed oil and meal are shown in **Figure 1**.

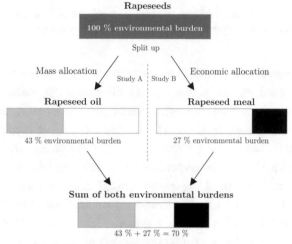

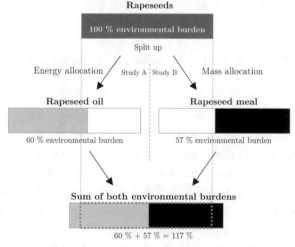

Figure 1 Splits and sum of the environmental burden of one agricultural system (Rapeseeds) when assessing by-products with different allocation approaches in independent LCA studies (study A for oil and study B for meal); Grey and black areas represent environmental burden
Upper part: Unintended ignoring of environmental burden due to different allocation approaches for by-products, with white space in the bar at the bottom illustrating ignored environmental burden; Lower part: Unintended double counting of environmental burden due to different allocation approaches for co-products, with dotted lines indicating initial environmental burden

(100%) and the overhanging bars visualizing double-counted environmental burden. The presented allocation shares originate from calculations and background data presented in Section 2.1.3.2.

The figure demonstrates the importance of the choice and consistency of allocation methods. Depending on the allocation approach, LCA practitioners might end up with diverging results for the same agricultural system (30% ignoring on one hand and 17% double counting of environmental burden on the other hand). In general terms, if independent LCAs use different allocation approaches for co-products that originate from the same agricultural system, the sum of the environmental interventions considered in each individual LCA is not equal to the actual environmental intervention of their common agricultural production process. This aspect should be carefully considered within the general interpretation of LCA results with any connection to agricultural processes.

2.1.1.4   The Cereal Unit as Basis for an Allocation Approach

With the goal of helping to address the allocation problem for agricultural LCAs, we identified the Cereal Unit (CU), a well-established unit in agricultural statistics in Germany. The CU makes possible the comparing of various agricultural products based on the animal feeding value. The animal feeding value is measured in repeatable and reproducible feeding experiments. Therefore, we consider the CU a physico-chemical and biophysical parameter.

A detailed description of the determination of the CU is given in the Supplementary data section. Because we applied no changes to the CU itself, the detailed description in the Supplementary data is only explanatory. The core of our work is Section 2.1.3. Here, we describe a new biophysical allocation approach that is based on the CU. Furthermore, we provide a comparison of the results from the CU allocation to those obtained via mass, energy and economic allocations.

2.1.2 Material and Methods — the Cereal Unit

The Cereal Unit (CU) was developed by German agricultural authorities and scientists decades ago [46]. Since 1944, it has been used and optimized continuously [46-50]. The CU is based on the animal feeding value, a relevant function of agricultural products. As a common denominator, the CU makes it possible to compare different agricultural products, including both vegetable and animal products.

Because publications on the calculation of the CU are out of print and, to the authors' knowledge, do not exist in the English language, we describe the scientific background of the calculation of the CU in the Supplementary data. For the application of the CU as a basis for allocation in LCAs (see Section 2.1.3), these calculations do not need to be performed; the existing exhaustive CU conversion tables can be used to perform CU allocation. The full list of CU conversion factors, which comprises over 200 products, is presented in the Supplementary data. Important indications about the geographical scope of the presented CU conversion factors in different regions are given in Section 2.1.4 and Section 2.1.5.

2.1.3 The Cereal Unit Allocation Procedure

Using several examples, we describe the implementation of the CU into an allocation procedure within this section.

2.1.3.1 Calculation Procedure

To perform the CU allocation procedure for a particular production process, the following steps are necessary.

1. All products and co-products of the process are identified.
2. The mass proportions of each involved product and co-product are known.

3. The CU conversion factors are determined for all products and co-products.

4. All product and co-product streams are converted into the CU.

5. The allocation ratio of each product and co-product is calculated using the CU.

A brief description of these steps and two supplementary examples are given in **Table 2**.

When applying the CU allocation procedure to wheat harvesting, the environmental burden of the relevant supply chain (growing a harvest-ready wheat plant) is allocated as follows: 75% to wheat grains and 25% to wheat straw. When applying the CU allocation procedure to the co-production of milk, live dairy cow and live calf for a one year observation period, the allocation shares are 86.6% to milk, 6.8% to live dairy cow and 6.6% to live fattening calf.

Table 2 Step-by-step instructions for the CU allocation procedure and the examples of wheat harvesting and the co-production of milk and live dairy cow and calf.

Step No.	Short description	Example 1	Example 2
1	Identify products and co-products	Process: Wheat harvesting; Product and co-product: wheat grains and wheat straw	Process: Co-production of milk and live dairy cow and calf (simplified); Product and co-products: the main product is milk, co-product 1 is live dairy cow, co-product 2 is live calf
2	Identify mass proportions	1 kg wheat plant → 0.56 kg wheat grains and 0.44 kg wheat straw	8 000 kg milk per year; 400 kg live-weight dairy cow with 5 lactations, meaning 400 kg / 5 years = 80 kg live-weight dairy cow per year; 180 kg live-weight fattening calf per year
3	Identify Cereal Unit conversion factors for products and co-products	1 kg wheat grains = 1.04 Cereal Units (CU); 1 kg wheat straw = 0.43 CU	1 kg milk = 0.80 CU; 1 kg live-weight dairy cow = 6.30 CU; 1 kg live-weight fattening calf = 2.69 CU
4	Convert product- and co-product-streams into Cereal Unit	0.56 kg * 1.04 CU/kg = 0.58 CU wheat grains; 0.44 kg * 0.43 CU/kg = 0.19 CU wheat straw	8000 kg * 0.80 CU/kg = 6400 CU milk; 80 kg * 6.30 CU/kg = 504 CU live-weight dairy cow; 180 kg * 2.69 CU/kg = 484 CU live-weight fattening calf
5	Calculate allocation ratio	0.58 CU / (0.58+0.19) CU equals 75% wheat grains; 0.19 CU / (0.58+0.19) CU equals 25% wheat straw	6400 CU / (6400+504+484) CU equals 86.6% milk; 504 CU / (6400+504+484) CU equals 6.8% live-weight dairy cow; 484 CU / (6400+504+484) CU equals 6.6% live-weight fattening calf

2.1.3.2   Comparison of Different Allocation Approaches

A comparison of existing allocation approaches for selected agricultural products from the cereal, sugar and oilseed sectors is discussed. Here the established allocation approaches, mass allocation (based on mass), energy allocation (based on a lower heating value) and economic allocation (based on market prices for January 2013), are compared to the newly introduced CU allocation. The following processes were considered:

- The harvesting process of wheat, leading to wheat grains and wheat straw;
- The flour milling process of wheat grains, leading to wheat flour and wheat bran;
- The harvesting process of rapeseed, leading to rape seeds and rapeseed straw;
- The oil milling process of rape seeds, leading to rapeseed oil and meal;
- The harvesting process of sugar beets, leading to sugar beets and sugar beet leaves.

The background data used for the calculation of allocation shares for the above-mentioned processes are listed in **Table 3**.

For **Table 3**, the mentioned data were, inter alia, obtained from [47, 51-58]. The allocation results are presented in **Table 4**.

Table 3   Background data for the calculation of allocation shares.

Co-products	Wheat harvesting		Flour milling of wheat grains		Rapeseed harvesting		Oil milling of Rape seeds		Sugar beet harvesting	
	Grains	Straw	Flour	Bran	Seeds	Straw	Oil	Meal	Beets	Leaves
Mass relation [kg]	0.56	0.44	0.86	0.14	0.37	0.63	0.43	0.57	0.59	0.41
Energy content [MJ/kg]	14.0	14.3	15.5	16.5	26.5	17.1	37.6	18.9	n.a.	n.a.
Economic value [€/t]	270	10	500	180	450	50	900	250	43.5	9.3
Cereal Unit conversion factor [CU/kg]	1.04	0.43	1.10	0.67	1.30	0.43	2.74	0.77	0.27	0.13

Table 4   Comparison of allocation results for agricultural products from the cereal, sugar and oilseed sectors.

Allocation type	Allocation share									
	Wheat		Wheat		Rapeseed		Rapeseed		Sugar beet	
	Grains	Straw	Flour	Bran	Seeds	Straw	Oil	Meal	Beets	Leaves
Mass	56%	44%	86%	14%	37%	63%	43%	57%	59%	41%
Energy	55%	45%	85%	15%	48%	52%	60%	40%	n.d.	n.d.
Economic	77%	23%	94%	6%	84%	16%	73%	27%	87%	13%
Cereal Unit	75%	25%	91%	9%	64%	36%	73%	27%	75%	25%

Compared to the established allocation approaches (mass, energy and economic), the new CU allocation approach leads to results that are in between those determined from other allocation approaches or even the same results; no extreme results were found.

2.1.4   Discussion

2.1.4.1   Cereal Unit Allocation versus Other Allocation Alternatives

Depending on the choice of allocation method, the functionalities of the co-products are expressed only to a limited extent. If applied to co-products used for energy production, the energy allocation might be more appropriate; however, this is not the case when one of the co-products from the same process does not have energy applications, such as sugar beet leaves and soybean meal. A comparable situation is given for mass allocation, which only renders a quantitative view of the co-products and ignores their inherent qualities (e.g., chemical properties for specific uses and nutrients for food or feed purposes). When economic allocation is applied, the distribution of the environmental burden between the co-products is oriented towards the prices that can be realized in the market. Potential price fluctuations would lead to different allocations of burdens between the co-products. Negative prices would lead to strange situations where negative environmental burdens would have to be addressed [12]. Furthermore, applying economic allocation in regions with different prices (e.g., caused by subsidies) would yield different allocation shares.

In addition, the mass content, energy content and, to some extent, economic value of a product are integrated in the CU because a certain amount of substance with a certain energy value (here metabolizable energy for animals) influences the feeding value and is mirrored by the CU. The economic value is represented via the share of the material in the feed mixture. For some co-products, the CU allocation results are very close or even identical to the economic allocation results. A potential reasoning behind this might be the typical use of those products as livestock feeds. In those situations, the economic value indicates the economic production potential when used as feeds. Economic allocation expresses this potential indirectly, whereas the CU immediately indicates this potential via the nutritional value.

Unique to a broad range of agricultural products is their use as livestock feed. Feed production, with an eighty percent share, is the largest user of agricultural area [59]. Applying animal feeding as a common-use perspective makes it possible to measure different agricultural co-products towards the same target.

Using the CU, the wide range of agricultural products can be compared based on their realistic utilization as livestock feeds. Therefore, CU allocation covers the potential uses of a broader range of agricultural co-products than established allocation approaches. The probability that different co-products are treated closer to their options of use in the real world might increase.

2.1.4.2 Vegetable and Animal in the Same System

The combination of different aspects and parameters reflected by the CU could be a promising compromise for depicting the intention of different users of agricultural products, which is relevant because agricultural production often includes different products, such as both vegetable and animal production. In many product systems, animal production and vegetable production are interlinked. Because the CU is valid for vegetable and animal products, LCA practitioners can use it to model agricultural farms as one system. An artificial split between crop farming and livestock farming is no longer necessary.

We applied CU allocation to the question of how to allocate the environmental burden of combined milk and meat production. Our results for CU allocation between milk (86.6%), dairy cow (6.8%) and fattening calf (6.6%) are very close to the cause-effect physical ('biological') allocation shares of 85% milk and 15% meat [60, 61]. Here, the "causal relationship between the dairy cow's feed mix and its production of milk, calves and meat" is calculated using Swedish fodder tables and Swedish feeding recommendations, resulting in a fix allocation of 85% for milk and 15% for meat and calves [60, 61].

Nguyen et al. [62] performed a very detailed consideration including six different feeding situations and a comparison of various allocation approaches, hereunder a biophysical allocation following the International Dairy Federation, the protein content and economic allocation. They found allocation shares between 74.0% and 89.5% for the milk [62]. The CU allocation results are very close to the calculation results of Cederberg and Mattsson [60] and Cederberg and Stadig [61] and in the same range of Nguyen et al. [62], even though different background calculation procedures took place.

In line with other authors, we think that dairy cows' feed should not be allocated between milk and calves only because a certain share of the feed energy is used for dairy cows' maintenance and, after a certain period of time, a dairy cow might be intended for meat production as well. Thus, a certain amount of the inputs should be allocated to the dairy cow. To obtain results for one year, we suggest using the reciprocal value of the number of lactations as a factor. The example mentioned before was based on five lactation periods. The given example with a number of only three lactations would be as follows: 82.8% milk, 6.3% live calf, 10.9% live dairy cow. The example shows that the CU makes it possible to include milk production performance (quantity of milk produced per year), the number of lactations, the intended uses of calves and the weights of dairy cows and calves.

CU conversion factors are available for live animals only. An additional allocation procedure is necessary for the slaughter process e for the division of meat, bones, hides, blood and further co-products. Additional industry-specific allocation approaches might be helpful for solving this additional allocation challenge. For the further processing needed to transform raw milk into dairy prod- ucts (e.g., butter, cheese and yoghurt), a industry-specific approach was developed by Feitz et al. [63]. Here, a dairy-specific physico-chemical allocation matrix was developed and could be used in combination with CU allocation. Both complement each other. The CU makes it possible to allocate to the point of producing raw milk, and the dairy-specific physico-chemical allocation matrix created by Feitz et al. [63] is suitable for the further processing needed to transform raw milk into the various dairy products.

2.1.4.3   Agricultural Co-Products in Supply-Chains

Co-products are often forced into the background in LCAs, especially if they are — at first glance — not immediately relevant for the analyzed process. However, one should take care of such co-product streams because they might have an influence on the overall result of the LCA. One example is cereal straw. In addition to traditional uses such as structure-lending components in feed material and for animal bedding, new processes using cereal straw as a raw material are emerging, such as in the production of biofuels [64-67] or as a feedstock for combustion plants [68]. Hereafter, the demand for straw can be expected to grow strongly. Because straw is obviously considered an agricultural product, it makes sense to allocate a certain amount of environmental burden to it, which is often not the case in LCAs because straw is classified as waste. The use of mass or energy allocation would end up allocating a relatively large allocation share to straw. This might explain why some LCA practitioners choose to allocate no environmental burden to the co-product straw. Allocating no environmental burden to one of the co-products in an LCA is seemingly inconsistent from a methodological point of view. Compared to mass or energy allocation, the CU allows practitioners to allocate a more substantiated amount of the environmental burden to straw, in particular, in smaller amounts and based on nutritive value (as a realistic use option for both co-products).

In regard to the choice of allocation procedure, LCA practitioners understandably often choose the allocation that best reflects the functionality of the co-product — e.g., energy allocation for bioenergy systems. In such cases, the functionalities of additional co-products might be underestimated, e.g., when used as feed. A solution to this problem could be choosing an allocation method that mirrors the functionalities of both the first and the second co-products. The chosen allocation method should be able to reflect functionalities that are covered by both co-products. Due to this relationship, such a choice of allocation is more substantiated by real usages compared to allocation choices where only the function of one co-product is considered. Because most agricultural co-products can serve as livestock feed and animal feeding value is expressed by the CU, this new CU allocation approach could be used for a very wide range of co-products.

2.1.4.4   Error and Uncertainty Analysis

Whereas mass, a lower heating value and market value can be directly measured with a known measuring error, the aggregation of different agricultural products contains, to some extent, a lack of sharpness that leads to inexactness of the results [69]. Because for the calculation of aggregated measures, several datasets are used and combined, they result in more than one source of uncertainty. Relevant for the CU are, e.g., the experimental measurement of the metabolizable energy for each livestock species on one hand and the estimation of actual feeding practices on the other hand. The estimations took place using the most recent and representative statistics and best available expert estimations. Due to a lack of a quantifiable measuring error for these estimations (i.e., the difference between the estimated value and true value is unknown) and a lack of provided information for error analysis, we cannot provide a quantitative analysis of errors that meets current scientific standards for error analyses.

The authors involved in updating the CU over the last decades continuously stressed the need for good quality estimations. The importance of this specific point has been known and documented throughout the development of the CU over the last decades. In this context, the wording "reasoned estimates" was established to express that all estimations must stand on a solid basis (e.g., official statistics) and expressing the awareness that the quality of their outcomes is directly affected hereby [47, 69-71]. The CU is widely known in agriculture. It is well established in German official agricultural statistics and has been continuously optimized over the decades [47, 49, 50].

2.1.4.5   The Cereal Unit — Representative of Most Agricultural Products and Co-Products

Independently of the actual use of agricultural products and co-products, CU allocation could be applied universally due to the following considerations. The majority of agricultural goods are suitable for feeding animals. In addition, "almost 80 percent of all agricultural land" is used to produce livestock feed [59]. Thus we consider livestock feed a major utilization path for agricultural goods. Because the majority of agricultural products are suitable for animal feed, a common-use perspective towards livestock feeding can be established, and a theoretical utilization as feed can be derived from this. The CU is based on the animal feeding value of agricultural products and serves to express the nutritional value to animals in one common unit.

Even in cases where agricultural products are not supposed to be used to feed livestock, the CU, as a common unit based on nutritional value, makes it possible to compare all involved agricultural products.

Agricultural products that cannot be used as feed or animal products are considered in the CU systematics as well. For those products, conversions and comparisons help to incorporate the entire range of agricultural products. The calculation approach for CU conversion factors for animal products is logically connected to the calculation for animal feed products. Products not intended for animal feeding (e.g., fruits, vegetables, tobacco and hop) are not measured based on their direct animal feeding value but are brought into relation to feedstuffs that could be grown on the same area with comparable yield, effort and revenue. Strictly speaking, this is a simplification of the calculation, but one must acknowledge that the largest share of agricultural area worldwide is used to produce animal feed; thus, the major user of agricultural products is well represented. Against this background, we deem this simplification to be within an acceptable range.

Although the CU is able to describe agricultural products that are not meant for livestock feed (e.g., fruits), the possibility of more detailed, sub-sector specific allocation approaches should be further investigated.

CU conversion factors are already available for more than 200 agricultural products and co-products (see chapter 2.2). The CU, a unit that integrates several features of different agricultural products and co-products, could potentially increase the probability of the same allocation method being used in two individual studies. This would avoid the phenomenon of unintended ignoring or double counting of environmental burden. Unless standardized, this phenomenon might not be completely avoided.

Conversion factors may not yet be available for a particular product, but they can be developed based on the published calculation method.

2.1.4.6   Cereal Unit in Other Countries

In a strict sense, the presented factors are valid only for Germany because the livestock composition and feed consumption of the region are part of the calculation of CU conversion factors. The CU is already well established in Germany as a commonly recognized unit for the calculation of official supply balances and economic accounts for agriculture, but other approaches have been developed from the following:

- Food and Agricultural Organization (FAO),
- Statistical Office of the European Union (EUROSTAT),
- Statistics Denmark,
- Department for Environment Food and Rural Affairs in the UK (Defra),
- French Central Office for Statistical Surveys and Studies (SCEES) together with the French CEREOPA-institute,
- Spanish Ministry of the Natural, Rural and Marine Environment (MARM),
- Italian Institute of Statistics (Istat) together with the Italian Institute for Agricultural Economics (INEA),
- Dutch Agricultural Economics Research Institute (LEI-DLO),
- Swiss Farmers' Union (SBV),
- United States Department of Agriculture (USDA).

These approaches show significant differences in their bases of calculation [48]; hence, they cannot be merged easily into one common unit that is
valid in all of these countries.

A globally valid system of conversion factors does not yet exist. However, the CU approach could easily be adapted to other countries. Geographical differences in animal-specific metabolizable energy rates, additional
feeding materials, additional livestock species and additional agricultural
products are relevant for the context in which the CU is to be applied. These
differences should be addressed in the calculation of CU conversion factors for
a specific region or even on the global level. In applying the concept of the
CU to globally traded agricultural products or co-products, a globally valid
list of CU conversion factors would be very important to avoid deviations
between different regions.

Formulas for the calculation of region-specific CU conversion factors are
given in the Supplementary data. The region-specific background information
necessary for these calculations often already exists, e.g., within agricultural
statistical reports and publications or can be obtained via expert consulta-
tion. Therefore, the CU approach theoretically does not have any geographical limitation. We do not see any substantial obstacles for the calculation of
conversion factors for specific regions, such as Asia, Europe and North and
South America, or even on a worldwide level. This step seems to be necessary
for the broader application of the CU.

The CU is updated in Germany approximately every 20 years. We did
not perform exhaustive investigations to define an updating interval, but, in
line with the recommendations from Becker [46] and given the considerable
effort needed for the calculations, we suggest updates every five or ten years.

2.1.5   Conclusions

The Cereal Unit (CU), an established measure in agricultural statistics, was
identified as a suitable parameter for agricultural allocation. In this study, a
new allocation procedure for agricultural LCAs based on the CU is proposed
and tested for selected products. From a theoretical point of view and sup-

ported by our results, the CU allocation approach offers vast potential; however, broad practical application is not yet available. To road test the proposed approach and either reconfirm its suitability or identify practical drawbacks, LCA practitioners are invited to use the CU allocation and publish their results.

While the data available for Germany are fairly complete, some data gaps in the list of conversion factors still need to be filled. A more substantial issue arises from the fact that the presented CU factors were derived for German conditions. While they may already serve as a proxy for Europe, other regions of the world with more significant differences in their agricultural production systems will need regional CU conversion factors. However, if the approach is used, this can be achieved without prohibitive effort. To calculate and test CU conversion factors for other regions, we intend to collaborate with experts from the agricultural sector of the other regions.

A potential application of an agriculture-specific allocation approach, such as CU allocation, may help reduce the variability and potential bias of LCA results in this sector. A reduction in the arbitrariness of LCA practitioners' decisions about allocation methods, on the one hand, decreases flexibility but, on the other hand, increases the reliability of agricultural LCAs and supports the use of LCAs by providing decision makers with more robust recommendations. Improved accountability, predictability and credibility for agricultural LCAs are the basis for new or improved effective strategies for the sustainable consumption and production (SCP) of products based on raw agricultural materials.

2.2 Supplementary Data – Application of the Cereal Unit in A New Allocation Procedure for Agricultural Life Cycle Assessments

This section contains the supplementary material, belonging to the publication:

Brankatschk, G., & Finkbeiner, M. (2014). Application of the Cereal Unit in a new allocation procedure for agricultural life cycle assessments. Journal of Cleaner Production, 73, 72-79. doi: http://dx.doi.org/10.1016/j.jclepro.2014.02.005

This supplementary data section comprises two main parts.

In a first part, Cereal Unit conversion factors are provided for more than 200 agricultural products and co-products, i.e. cereals (tab 1), oilseeds (tab 2), vegetable oils and fats (tab 3), products from cereal processing (tab 4), products from oilseed processing (tab 5), products from starch processing (tab 6), products from distilleries and breweries (tab 7), products from sugar beet and fruit processing (tab 8), roots and tubers (tab 9), roughage (tab 10), products from food production (tab 11), feedstuffs of animal origin (tab 12), specialty crops (tab 13), fruits (tab 14), vegetables I (tab 15), vegetables II (tab 16) and products from livestock farming (tab 17). This collection of Cereal Unit conversion factors allows immediate use of this new allocation approach and makes the Cereal Unit allocation approach easy to use for LCA practitioners (*contributing to research targets #4,10*).

The second part provides an historical summary and a detailed description of the calculation procedure of the Cereal Unit, which is mainly based on the metabolizable energy and used since decades in official German agricultural statistics. This is the first English translation of the entire calculation procedure for the calculation of Cereal Unit conversion factors and the first English list of conversion factors. These detailed explanations pave the way

for the calculation of additional Cereal Unit conversion factors for further
agricultural products and for further regions (*#3,4,10*).

2.2.1   Cereal Unit Conversion Factors for Several Agricultural Products

Cereal unit conversion factors for several agricultural products and co-
products, presented in the following tables (**Table 5 — Table 21**) were
derived from Mönking et al. [47].

Reading examples:

     1.00 kilogram **Feed wheat** = **1.04** kilogram Cereal Unit
     0.96 kilogram Feed wheat = 1.00 kilogram Cereal Unit

     1.00 kilogram **Feed barley** = **1.00** kilogram Cereal Unit
     1.00 kilogram Feed barley = 1.00 kilogram Cereal Unit

     1.00 kilogram **Feeding oats** = **0.84** kilogram Cereal Unit
     1.19 kilogram Feeding oats = 1.00 kilogram Cereal Unit

Table 5   Cereal Unit conversion factors for cereals

Feed wheat	1.04
Feed barley	1
Feed rye	1
Feeding oats	0.84
Triticale	1
Maize corn	1.08
Corn cob mix, CCM	0.71
Fodder Maize, whole plant	1.1
Feed millet	0.86
Feed rice	0.82

Table 6   Cereal Unit conversion factors for oilseeds

Soybeans	1.15
Rape seeds	1.30
Sunflower seeds	1.25
Cotton seeds	0.96
Lin seeds	1.21

Table 7   Cereal Unit conversion factors for vegetable oils and fats

Soybean oil	2.81
Rapeseed oil	2.74
Groundnut oil	2.67
Palmkernel oil	2.26
Coconut oil	2.68

Table 8   Cereal Unit conversion factors for products from cereal processing

Bran	
from Wheat	0.67
from Barley	0.77
from Rye	0.67
from Oat	0.83
from Maize corn	0.88
Semolina bran	
from Wheat	0.75
from Rye	0.72
Middlings / Feed meal	
from Wheat	0.89
from Barley	0.81
from Rye	0.86
from Oat	1.13
from Maize corn	0.95
Screenings	
from Wheat	0.98
from Barley	0.00
from Rye	0.95
from Oat	0.00
from Maize corn	0.93
Chaff / Husks	
From Oat	0.54
Rolled grains / flakes	
from Oat	1.02
from Maize corn	0.97
from potato	0.85

Table 9   Cereal Unit conversion factors for products from oilseed processing

Oilmeals	
from Soybeans	0.96
from Rape seeds	0.77
from Sunflower seed	0.75
from Lin seeds	0.84
from Sesame seeds	0.85
from Groundnuts	0.96
from Coconuts	0.88
from Cotton seeds	0.78
from Palm kernels	0.71
from Maize germs	0.89
Oil expeller	
from Rape seeds	0.89
from Sunflower seed	0.80
from Lin seeds	0.89
from Sesame seeds	0.95
from Groundnuts	0.91
from Coconuts	0.90
from Cotton seeds	0.85
from Palm kernels	0.89
Oil cake	
from Rape seeds	0.94
from Sunflower seed	0.84
from Lin seeds	0.94
from Sesame seeds	0.88
from Groundnuts	1.02
from Coconuts	0.93
from Cotton seeds	0.90
from Palm kernels	0.94
from Maize corn	0.92
Soybean hulls	0.69

Table 10 Cereal Unit conversion factors for products from starch processing

Maize starch	1.07
Maize gluten	1.22
Maize gluten feed	0.82
Maize steeping liquor	0.50
Potato starch	0.98
Potato protein	1.28
Potato pulp	0.78
Wheat gluten	1.25
Wheat gluten feed	0.88

Table 11 Cereal Unit conversion factors for products from distilleries and breweries

Floating barley grains	0.48
Malt sprouts / Malt germs	0.74
Malt spent grains / Brewers' spent grains / draff	0.75
Brewers' yeast	0.91
Distillery spent wash from potato	0.05
Distillery spent wash from wheat	0.06
Distillery spent wash from rye	0.06
Distillery spent wash from maize	0.08

Table 12 Cereal Unit conversion factors for products from sugar beet and fruit processing

Citrus pomace / pulp	0.88
Fruit pomace / pulp	0.73
Grape pomace / pulp	0.38
Sugar cane molasses	0.75
Sugar beet molasses	0.79
Molassed sugar beet chips	0.87
Unmolassed dried sugar beet chips	0.81

Table 13 Cereal Unit conversion factors for Roots and tubers

Sugar beet	0.23
Fodder beet (10% saccharose)	0.14
Fodder beet (4% saccharose)	0.11
(late) Turnip	0.09
Potato	0.22
Cassava / manioc	1.03
Topinambour / Jerusalem artichoke	0.85

Table 14 Cereal Unit conversion factors for roughage

Cereal straw	0.43
Grass, fresh	0.16
Silage from Grass	0.27
Hay from Grass	0.61
Maize silage	0.30
Maize, fresh	0.30
Sugar beet leaves	0.13
Fodder beet leaves	0.15
Catch crop forage rape	0.10

Table 15 Cereal Unit conversion factors for products from food production

Old bread	0.78
Bread waste	0.93
Biscuits	1.34
Pasta waste	1.02
Leftover foodstuffs	0.23

Table 16 Cereal Unit conversion factors for feedstuffs of animal origin

Unskimmed milk; for animal feeding	0.22
Goat milk	0.21
Skimmed milk	0.10
Buttermilk	0.03
Whey, sweet	0.07
Whey, acid	0.06
Unskimmed milk powder	1.46
Whey powder	0.90
Sweet whey powder	0.99
Skimmed milk powder	1.10
Buttermilk powder	1.09
Casein	1.43
Animal body meal	1.02
Meat meal	1.48
Meat and bone meal	0.68
Bone meal	0.58
Bone meal, degelatinised	0.08
Blood meal	1.15
Fish meal	1.17
Fish solubles	0.77
Animal fats	2.75

Table 17 Cereal Unit conversion factors for specialty crops

Fruits	0.56
Tree fruits	0.47
Bush fruits	2.65
Strawberries	1.16
Vegetables	0.42
Cabbage vegetables	0.31
Leaf vegetables	0.61
Stem vegetables	1.42
Root and tuber vegetables	0.30
Fruiting vegetables	0.41
Legumes	0.81
Other vegetables	0.35
Wine	1.32
Red wine	1.39
White wine	1.22
Hop	6.85
Tobacco	3.26
Hemp	1.00
Fiber flax	1.23
Grass seeds	5.00
Clover seeds	10.00
Medicinal and Aromatic plants	0.38

Table 18 Cereal Unit conversion factors for fruits

Tree fruits	0.47
Apples	0.42
Pears	0.77
Cherries, sweet	2.73
Cherries, sour	2.27
Plums and Damsons	1.46
Mirabelles and Greengages	1.51
Apricots	4.59
Peaches	2.94
Walnuts	5.73
Bush fruits	2.65
Currants (black, red, white)	2.62
Gooseberries	n. d.
Raspberries	2.34
Blueberries	3.42
Sea buckthorn	5.45
Strawberries	1.16

Table 19 Cereal Unit conversion factors for vegetables I

Cabbage vegetables	0.31
Cauliflower	0.49
Broccoli	0.87
Chinese cabbage	0.32
Kale	0.73
Kohlrabi	0.45
Brussels sprouts	0.79
Red cabbage	0.23
Cabbage	0.18
Savoy cabbage	0.38
Leaf vegetables	0.61
Oak leaf lettuce	0.51
Iceberg lettuce	0.49
Endive	0.43
Corn salad	0.89
Lettuce	0.46
Lollo lettuce	0.52
Radicchio	0.61
Romaine lettuce	0.68
Rocket / Arugula / Rucola salad	0.59
Other salads	0.42
Spinach	0.74
Stem vegetables	1.42
Rhubarb	0.55
Asparagus	1.71
Stalk celery	0.37

Table 20 Cereal Unit conversion factors for vegetables II

Root and tuber vegetables	0.30
Florence fennel	0.50
Knob celery	0.32
Horse radish	1.16
Carrot	0.25
Small / European radish	0.47
Radish	0.44
Beetroot / Red beet	0.32
Fruiting vegetables	0.41
Gherkins / Pickled cucumbers	0.21
Cucumber	0.41
Squashes	0.44
Zucchini / Courgette	0.35
Sweet corn	0.53
Legumes	0.81
Bush beans	0.39
Broad beans	0.54
Runner beans	0.23
Peas, fresh, without pod	0.79
Peas, fresh with pod	0.47
Other vegetables	0.35
Scallion / Spring onion	0.37
Common onion	0.29
Parsley	0.43
Leek	0.38
Chives	0.43
Other vegetables	0.23

Table 21 Cereal Unit conversion factors for products from livestock farming

Milk, unskimmed, for human nutrition	0.80
Live weight	
Cattle, total	5.98
Calf, at birth	4.52
Calf, intended for rearing	3.49
Heifer	6.27
Dairy cow	6.30
Fattening bull	5.70
Fattening ox	6.33
Fattening heifer	6.53
Fattening calf	2.69
Live weight	
Pork	3.05
Sow	2.87
Porker	3.06
Eggs	2.28
Live weight	
Poultry	2.55
Chicken	2.27
Broiler	2.07
Layer / Laying hen	4.60
Turkey	3.06
Duck	2.48
Goose	5.10
Live weight	
Sheep	9.1
Goat	2.91
Wool, raw	1.9
Goat milk	0.78
Live weight	
Rabbit	4.45
Intensive husbandry	2.38
Extensive husbandry	4.93

2.2.2   Material and Methods – The Cereal Unit

Historically, different estimation approaches for agricultural productivity
have been discussed [46]. A simple summation of the masses of all agricultural products was recognized as being an inappropriate solution because
different levels of effort for the production, the various functionalities of the
products and fulfillment of services are neglected. To make agricultural
productivity more comparable and measurable, it was realized that an
aggregation in terms of a weighted sum, rather than a summation of the
masses, is necessary [46]. In contrast to the simple summation of masses, a
weighted sum or aggregation takes into account the functionalities of each of
the products. Here, the challenge of finding an appropriate way to mirror the
functionalities of the agricultural products and co-products emerged. "Almost 80 percent of all agricultural land" is used to produce livestock feed
[59]. This land share might also be transferred to the consumption of agricultural products. In addition, a very large share of all agricultural goods is
suitable for feeding animals. Therefore, the animal feeding value is well
suited as a key parameter to express the functionality of agricultural products and co-products. The animal feeding value is the main basis for the
Cereal Unit. It was developed by German agricultural authorities and
scientists decades ago [46]. Since 1944, it has been used and optimized continuously [46-49, 72]. The Cereal Unit is based on a major functionality –
the animal feeding value – and can be used as a common denominator; it even
makes it possible to compare different products to one another. Solutions to
expressing the animal-specific metabolization rates of different agricultural
products and describing agricultural products that cannot serve as feed were
found during the development of the Cereal Unit in the last decades. Within
the following sub-sections, the scientific basis for the calculation of Cereal
Unit is described in detail.

2.2.2.1 Calculation of the Cereal Unit

The main focus of the Cereal Unit is the animal feeding value of agricultural products and co-products (including both vegetable and animal products). Primarily, the Cereal Unit is calculated for vegetable products intended as livestock feed (2.2.2.1.1). Only in cases where the agricultural product does not serve as feed are complementary calculation methods for vegetable products (2.2.2.1.2) or animal products (2.2.2.1.3) applied. In theory, Cereal Unit conversion factors can be calculated for all agricultural products. Publications on the calculation of the Cereal Unit are out of print and, to the authors' knowledge, do not exist in the English language. Therefore, to ensure comprehensibility, we briefly present the calculation for the Cereal Unit. The description of the calculation of the Cereal Unit is based on Mönking et al. [47], with slight modifications. For the application of the Cereal Unit as a basis for allocation in LCA practices, the existing exhaustive conversion tables might be used, which are based on the processes described in sections 2.2.2.1.1, 2.2.2.1.2 and 2.2.2.1.3. Some selected Cereal Unit conversion factors are presented in section 2.2.2.1.4.

2.2.2.1.1 Cereal Unit of Products Used as Livestock Feed

To express the total energy content of an agricultural product, the gross energy content (GE) is used (see **Formula 1**). The energy contents of the following macronutrients are considered: crude protein (CP), crude lipids (CL), crude fiber (CF) and nitrogen-free extracts (NFE; containing hydrocarbons). The definitions of these macronutrients are in line with common feed material analysis, the so-called Weende Analysis or proximate analysis [73].

$GE\ [MJ] = 23.9 * CP\ [kg] + 39.8 * CL\ [kg] + 20.1 * CF\ [kg] + 17.5 * NFE\ [kg]$

Formula 1 Calculation of the gross energy by adding together the energy contents of products' macronutrients. GE = gross energy in megajoules (MJ); CP = crude protein; CL = crude lipids; CF = crude fiber; NFE = nitrogen-free extracts; [kg] = Amount reported in kilograms [47, 73].

Because of the different digestive systems of livestock species and the resulting different use efficiencies for different feed contents, only a certain share of the gross energy content can be metabolized. This is expressed by different metabolizable energy contents for each feed material and each animal species. Because feed composition, breed-specific metabolization rates and the assessment of metabolization rates itself underlie variations, the metabolizable energies are not fixed and underlie fluctuations. These variations are not described here, but the calculations, on which the Cereal Unit is built, are supposed to be based on the most representative and recent data.

The data used by Mönking et al. [47] to calculate the recent version of the Cereal Unit were obtained from official and published feed composition statistics such as the 1970, 1984, 1997 and 2009 editions of the feed databases of the German Agricultural Society (Deutsche Landwirtschaftsgesellschaft, DLG), the 2004 edition of the nutritional value tables from the French INRA institute and the 2007 edition of the Netherlands' CVB tables [cited in 47]. Examples for the average Metabolizabe Energy values that have been used in the latest Cereal Unit update are given in **Table 22**.

Table 22 Examples for Gross Energy and Animal-specific Metabolizable Energy contents for selected feed materials [47]

Agricultural products	Gross Energy [MJ/kg fresh mass]	Metabolizable Energy content [MJ/kg fresh mass]			
		Cattle	Pigs	Poultry	Horses
Straw	15.29	5.16	3.1	1.35	5.58
Hay	15.56	7.72	5.7	5.5	7.22
Wheat	16.41	11.77	13.79	12.78	12.17
Barley	16.28	11.30	12.63	11.43	11.65

To understand the overall feeding value of a certain agricultural product, it is very time consuming to look at the animal-species specific metabolizable energy rates. As a proxy, a combination of the animal-specific metabolization rate (**Table 22**) and the real usage quantity (**Table 23**) can be used as an aggregation parameter. The proxy is called the Specifically aggregated metabolizable energy content (**Table 24**).

Mönking et al. [47] identified the share of feed materials fed to specific animal species for Germany (see **Table 23**). This information can be obtained via expert estimation, standard feed formulations and surveys. Such data can be found for each region, such Europe and North America, or even on a global level. The idea of expressing the feeding value in feed units is a well-known and established approach in animal nutrition. Detailed descriptions of some of these approaches are available [74].

Table 23 Share of feed material fed to different animal species in Germany [47]

Agricultural products	Used to feed ... [mass %]			
	Cattle	Pigs	Poultry	Horses
Wheat	20	54.5	25	0.5
Barley	5	94.9	0	0.1

The share of feed material used (**Table 23**) and the metabolizable energy (**Table 22**) are used to calculate the specifically aggregated metabolizable energy content for each animal feed. An example is given in **Table 24** for barley.

Table 24 Calculation of specifically aggregated metabolizable energy content for barley [47]

	Metabolizable Energy [MJ / kg fresh mass of barley] – data from **Table 22**	Share of feed material use [%] – data from **Table 23**
Cattle	11.3	5
Pigs	12.63	94.9
Poultry	11.43	0
Horses	11.84	0.1
Specifically aggregated metabolizable energy content	12.56	100

The specifically aggregated Metabolizable Energy content of 1 kilogram of barley, namely 12.56 MJ, has been defined as 1 *Cereal Unit* (in German "Getreideeinheit"). Hence, in earlier times, the Cereal Unit was called a barley unit (in German "Gersteneinheit") [46]. The specifically aggregated Metabolizable Energy contents of other feed materials are brought in relation to this reference value for barley. In this way, Cereal Unit conversion factors have been calculated for a large number of agricultural products. A short selection is given in section 2.2.2.1.4.

2.2.2.1.2   Cereal Unit of Vegetable Products Not Used as Livestock Feed

There are agricultural products that are not assessed as livestock feed. These so-called specialty crops are not intended for animal feeding. Examples include fruits, vegetables, herbs, tobacco, hop and flowers. To include them in the Cereal Unit system, an auxiliary calculation approach has been developed over decades [46, 47]. The specialty crops are brought into relationship to one of the three intensity levels of reference crops. Those reference crops are defined based on their average yield and are expressed in Cereal Units per hectare (see **Table 25**).

The decision as to which intensity levels specialty crops belong is based on a comparison to the reference cultures based on the agronomic and economic aspects of the growth of these specialty crops. For example, soil conditions, climatic conditions, crop rotation and the (work) intensity of cultivation are relevant aspects; an extensive description is provided in Becker [46]. In a nutshell, the specialty crop is assigned to the intensity level in which, e.g., the workload, agronomic production conditions, economic production potential and yield of the specialty crop are comparable. For example, strawberry production is very labor intensive, allows good margins and is comparable in terms of additional aspects to the reference crops that represent intensity level 1. Therefore, strawberries are assigned to intensity level 1.

For the calculation of the Cereal Unit conversion factor of the specialty crop, the yield of the reference culture is brought into relation to the yield of the specialty crop (see **Formula 2**).

Table 25 Intensity levels for the valuation of specialty crops [47]

Intensity level	Yield of reference culture [CU/ha]	Examples
1	13 000	Beets, silage maize
2	8 000	Wheat, corn maize, potatoes
3	4 000	Legumes, modest cereals

Cereal Unit conversion factor of specialty crop [CU/kg]
= yield of reference culture [CU/ha] / yield of specialty crop [kg/ha]
Formula 2 Cereal Unit conversion factor calculation for specialty crops (vegetable products not intended as livestock feed) [47]

As an example, we demonstrate the calculation of the Cereal Unit conversion factor for strawberries. Based on the agricultural statistics, the average yield of strawberries is 11 250 kg/ha. Because strawberries belong to intensity level 1, the yield of the reference culture (13 000 kg CU/ha) is divided by 11 250 kg/ha, leading to the Cereal Unit conversion factor for strawberries being 1.16 Cereal Units per kilogram of strawberries [47]. The Cereal Unit can also be calculated for groups of specialty products. Here, the harvest share of the sub-product belongs to the product group used for the specific aggregation procedure (see **Table 26**).

Table 26 Calculation of Cereal Unit conversion factors for groups of specialty crops – Example of wine [47]

Specialty crop	Cereal Unit conversion factor per kilogram of product [CU/kg]	Specific aggregation factor = Share of total harvest [%]
White wine	1.39	60.16
Red wine	1.22	39.84
Wine, total	1.32	100

2.2.2.1.3   Animal Products

Cereal Unit conversion factors are available for animal products as well. In this case, the energy content of the animal product itself is not the decisive factor. The calculation is based on the amount of animal feed that is necessary to produce this specific animal product [47]; see **Formula 3** as well.

Cereal Unit conversion factor of animal product
= feed energy demand per kg product in [MJ metabolizable energy] / animal
specific energy content of 1 kg barley [MJ metabolizable energy]
Formula 3 Cereal Unit conversion factor calculation for animal products [47]

An example for the calculation of the Cereal Unit conversion factor for 1 kg of milk is given in **Formula 4**. To express 1 kg of milk in the system of Cereal Units, the feed demand for the production of 1 kg of milk is divided by the Metabolizable Energy content of barley; in particular, 1 kilogram of milk equals 0.8 Cereal Units.

Cereal Unit conversion factor 1 kg milk
= feed demand for production of 1 kg milk [MJ Metabolizable Energy] / cattle
specific energy content of barley [MJ Metabolizable Energy]
= 9 MJ Metabolizable Energy / 11.3 MJ Metabolizable Energy
= 0.8 CU/kg milk
Formula 4 Calculation of Cereal Unit conversion factor for 1 kilogram of milk [47]

2.2.2.1.4   Summary of Cereal Unit Calculation

As described in sections 2.2.2.1.1, 2.2.2.1.2 and 2.2.2.1.3 the Cereal Unit uses several science-based elements and parameters that are relevant for agricultural products to express the value of a particular agricultural product in relation to barley, which serves as a reference product. A list of conversion factors is shown in **Table 27** and **Table 28** for selected field crops and several corresponding co-products. The full list of Cereal Unit conversion factors, comprising over 200 products is presented in section 2.2.1.

The Cereal Unit conversion factor of 1.00 for barley grain expresses the animal feeding value of 1 kg of barley grains. The Cereal Unit conversion factor for each agricultural product refers to 1 kg of the respective product and gives an indication about the animal feeding value of each of the products normalized to barley. For example, the Cereal Unit conversion factor of 1.04 for wheat grain means a 4 per cent higher value coming from 1 kg of wheat compared to 1 kg of barley. The Cereal Unit conversion factor of 0.43 for Cereal straw indicates a distinctively lower nutritive value for 1 kg cereal straw compared to 1 kg barley grains.

Table 27 Selected Cereal Unit conversion factors for agricultural products and co- products; for German conditions – part 1 cereals [47, 49, 72]

Field crop	Cereal Unit conversion factor [CU / kg product]	Co-product	Cereal Unit conversion factor [CU / kg product]
Cereals			
		Cereal straw (without distinction between types of cereals)	0.43
Barley	1.00	Malt sprouts	0.74
		Malt spent grains / brewers' spent grains / draff	0.75
		Beer yeast	0.91
Wheat	1.04	Distillery spent wash from wheat	0.06
Rye	1.01	Distillery spent wash from rye	0.06
Oat	0.84		
Triticale	1.01		
Maize corn	1.08	Maize germ meal	0.89
		Corn gluten feed	0.82
		Distillery spent wash from maize	0.08
Millet	0.86		
Rice	0.82		
		Bran	0.72

Table 28 Selected Cereal Unit conversion factors for agricultural products and co- products; for German conditions – part 2 oilseeds, roots and tubers [47, 49, 72]

Field crop	Cereal Unit conversion factor [CU / kg product]	Co-product	Cereal Unit conversion factor [CU / kg product]
Oilseeds			
Rape seeds / canola seeds	1.30	Rapeseed oil	2.74
		Rapeseed meal	0.77
Soybeans	1.15	Soybean oil	2.81
		Soybean meal	0.96
Sunflower seeds	1.25	Sunflower meal	0.75
Linseeds	1.21	Linseed meal	0.84
Cotton seeds	0.96	Cottonseed meal	0.78
		Palm kernel oil	2.26
		Palm kernel meal	0.71
Roots and Tubers			
Potato	0.22	Distillery spent wash from potato	0.05
Sugar beet	0.23	Sugar beet leaves	0.13
		Sugar beet molasses	0.79
		Molassed sugar beet chips	0.87
		Unmolassed dried sugar beet chips	0.81
Fodder beet	0.14	Fodder beet leaves	0.15
		Sugar cane molasses	0.75
Cassava / manioc	1.03		
Topinambour	0.85		
Roughage			
Grass, fresh	0.16		
Grass silage	0.27		
Grass hay	0.61		
Maize silage	0.30		
Catch crop forage rape	0.10		

2.3   Modeling Crop Rotation in Agricultural LCAs — Challenges and Potential Solutions

This section contains the publication:

Brankatschk, G., & Finkbeiner, M. (2015). Modeling in agricultural LCAs — Challenges and potential solutions. Agricultural Systems, 138, 66-76. doi: http://dx.doi.org/10.1016/j.agsy.2015.05.008

Within this research article, crop rotation effects between agricultural crops, grown in temporal succession on the same field are indicated and the need to consider them in environmental assessments is explained. Current LCA methodology is described having limited ability to cover all physical, chemical and biological properties of agricultural land and improved phytosanitary conditions. A new approach for the modeling of crop rotations is developed and presented step by step. First stage of this crop rotation approach is to extend the system boundary in order to include the entire crop rotation. This procedure achieves including long-term farming strategies and phytosanitary effects amongst crop rotation elements (*contributing to research targets #5, #6 and #7*). In a second stage, the environmental burdens of entire crop rotation are allocated among all products and co-products, using an agriculture-specific allocation approach, e.g. the Cereal Unit allocation approach (*#3*). This procedure allows evaluating modified crop rotations, whilst keeping the product focus when assessing different crop rotations (*#7, 8*). The new crop rotation approach is compatible to the existing life cycle inventory (LCI) as one part of the ISO standardized LCA procedure and other LCA-steps, i.e. Goal and scope definition, life cycle impact assessment (LCIA) and interpretation remain unaffected (*#10*). In a brief example for applying all necessary steps of the new crop rotation approach, numerical differences in LCI results are provided (*#9*).

Abstract

Despite large efforts there are still methodological challenges to bring life cycle modeling closer to agricultural reality. Here, we focus on the inclusion of the effects occurring between the crops grown in the same agricultural field in temporal succession. These so called crop-rotation effects are caused by changes in physical, chemical and biological properties of the agricultural land over time (presence and availability of different micro and macronutrients, soil structure, soil texture, phytosanitary conditions, presence of weeds, etc.) due to the rotation of crops. Since a huge number of parameters contribute to crop-rotation effects, they cannot be easily measured. Therefore, LCA (Life Cycle Assessment) studies with system boundaries containing only one vegetation period have a limited ability to include these effects — unless explicit modeling measures have been taken to include individual crop-rotation effects. Existing approaches for the inclusion of crop-rotation effects are described, e.g. via transferring certain amounts of nutrients and their environmental burdens to subsequent crops. Still, many crop-rotation effects between crops are not covered in recent LCA methodology; corresponding gaps are identified and described. Examples include reduced input of agrochemicals via improved phytosanitary conditions, stabilization of yields via reduction of harvest failures, improved yields via improved soil texture, soil structure and improved conditions for soil organisms. Overall, most crop-rotation effects are not properly addressed in current LCA practice. Thus, LCA results and the quality of derived recommendations are negatively affected — for example incentives for the (unlimited) removal of crop residues are set based on LCA results without considering potential adverse effects on soil fertility. In other words, these gaps might lead to unintended free-rider problems. A new approach for the modeling of crop-rotation effects is suggested. It consists of six steps. First, align the system boundary during the inventory analysis to the level of the whole crop rotation system; second, determine all inputs of the whole crop rotation; third, do the same for the outputs; fourth, convert all outputs to a common agriculture-specific denominator, the so-called Cereal Unit; fifth, calculate an output-specific allocation share using the ratio of each individual output to the sum of all outputs of the crop rotation; and sixth, apply the allocation shares to the sum of each input-type — resulting in the output-specific allocated input. One major advantage of this approach is the integration of crop-rotation systems into LCA, including all relationships between the individual crops of the crop rotation. Using this approach, LCA practice becomes able to depict crop rotations more accurately and to avoid the current practice of ignoring the effects between individual crops. It might enable LCA to consider the fundamental agricultural principle of crop rotations and to include interactions between one crop and the subsequent crop. Since these crop-rotation effects influence soil fertility, yields and overall sustainability of agricultural systems, the reliability of the evaluation of environmental impacts might be affected. Thus, the ability to consider the entire spectrum of crop rotation effects should be integrated into agricultural LCAs.

2.3.1   Introduction and Problem Description

Evaluation of environmental burdens caused by the production of goods or the provision of services, including their agricultural supply chains, is well established in Life Cycle Assessment (LCA). Several articles describe future challenges and potential suggestions for further methodological improvement of agricultural LCA methodology in general [75-77], for food and feed [78, 79], renewable materials [80] and bioenergy systems [29]. Further approaches for methodological improvement might be found in the proceedings of the LCA Food conferences [3-7, 81].

Even though much effort has been invested in improving agricultural LCA, there are still methodological challenges, e.g. the allocation between co-products and the consideration of crop rotations. Agricultural production systems and the processing of their products typically lead to multiple outputs, e.g. straw and grain from cereal harvesting, flour and bran from grain milling, vegetable oil and oilseed meal from oilseed crushing. The proper attribution of environmental burdens between individual co-products is one major methodological challenge. A number of publications dealing with this topic and offer several solution approaches [11, 15, 33-36, 38, 40].

Another challenge for proper representation of agricultural reality is the consideration of crop rotations in LCA. "If an LCA study focuses on just one crop [...], it fails to account for the interactions between this crop and preceding and subsequent crops" [82]. Agricultural systems are highly complex, and their functional principles are quite well understood, but not all underlying material flows can be easily quantified. To have a complete picture of involved substances and material flows, it is essential to find system boundaries that are equally valid, both in agricultural practice and in the LCA model. This is relevant, because the quality of this representation affects the quality and meaningfulness of the overall LCA results. Finding appropriate system boundaries might be relatively easy for chemical reactions – if they take place in test tubes or systems with clear physical borders – but such clear borders do not exist for agricultural systems.

In agricultural LCAs, typically, one vegetation period is used as the system boundary, and thus often only one crop, from seedbed preparation or sowing to harvesting, is included. In this case, the definition of the system boundary around the studied crop is time-oriented. The influence of the previous crop on the assessed crop in the same field is thus outside the scope — unless explicit measures have been taken to include it, e.g. by accounting for nutrient transfer from one crop to the subsequent crop. Considering just one vegetation period is a source of errors in agricultural LCAs, since the nutrient supply, one relevant contributor to environmental interventions, is affected. Each agricultural crop influences the nutrient content in the soil; the amount of nutrients in the soil at the beginning of each vegetation period might not be equal to that at the end of this period. In good agricultural practice, fertilization activities depend on the nutrient content in the soil and attempt to leave the soil with an optimum amount and balance of nutrients. By disregarding this rationale behind the nutrient balances, incorrect assumptions about nutrient consumption could be drawn — leading to imprecise attribution of nutrient consumption to the individual crops.

Another example is the improvement of phytosanitary conditions that comes from changing the crops grown on the same field in temporal succession, because different parasites and diseases are linked to different types of crops, and some crops serve as cultivation breaks or even suppress some infectious agents. Since crop-rotation effects affect, for example, the presence and availability of nutrients, soil structure, timing of farming activities and pest control by changing crops, they influence the achievable yield and environmental impacts of each individual crop that is grown in crop rotations.

The following sections provide a short historical outline of the evolution of understanding crop-rotation effects (Section 2.3.1.1); furthermore, versatile benefits of crop rotations are explained (Section 2.3.1.2). Several examples of crop-rotation effects are mentioned (Section 2.3.1.3), and different approaches are described for considering them in LCAs (Section 2.1.1.4). None of them covers the full range of crop-rotation effects and thus none is completely satisfactory. So far, no agreement has been achieved about whether and how the various crop-rotation effects are to be included in LCA via a uniform approach. The crop-rotation effects are physically real, described in agricultural publications [83-85] and do have strong influence on agricultural practices e.g. cultivation planning, plant protection and plant nutrition. Practical needs for including these effects in LCA are mentioned in Section 2.3.1.5.

2.3.1.1   Historical Outline of Crop Rotation

"The object of the art of agriculture is to make the soil permanently yield the largest possible quantity of valuable produce in the shortest possible period of time. [...] As the most valuable crops always [...] diminish the future fertility of the land, measures must be adopted for restoring that fertility by other crops" [86]. The fact that different crops have different demands on the soil (e.g. nutrient demands) and different effects on soil fertility and yields has been well known for centuries.

Since the middle of the 19th century it has been known that the plant-specific uptake of nutrients influences the type and amount of nutrients in the soil after the crop and thus affects the yield of the subsequent crop grown in the same field [87, 88]. The importance of the availability of nutrients, along with the limitation of plant growth by the scarcest resource, was described by Liebig et al. [88] and is today widely known in agronomy as "Liebig's Law of the Minimum".

Even without a comprehensive understanding of underlying physical relationships, farmers developed appropriate agricultural practices centuries ago to avoid nutrient deficiency and increase yields, e.g. by developing the so-called three-field system. The three-field system is a well-known historical example of a farming system from the Middle Ages and indicates the importance of crop-rotation effects [89, 90].

2.3.1.2 Features of Crop Rotations

Crop rotation describes the sequence of different agricultural crops grown on the same field. In growing different crops in chronological sequence, positive effects from the current to the subsequent crop can be achieved [91]. For example, improvement of phytosanitary conditions reduces disease pressure and infestation by parasites. The reasoning behind is the change in crops that creates a time gap in which no host is available for crop-specific parasites or diseases. Another example is the improvement in nutrient availability of the subsequent crop. Here the crops are using different nutrients or leaving different nutrients in residues or sourcing the nutrients from different soil horizons. These crop-rotation effects can be physically measured by long-term field experiments and are well described in scientific publications [84, 85, 92]; they are important for agricultural practice — e.g. in terms of crop planning and supply of nutrients to plants [86, 90, 93, 94]. Agronomists and soil scientists see a clear relationship between crop rotations and sustainability of agricultural production systems [83, 95, 96].

Several interactions lead to positive crop-rotation effects. Cowell et al. [82] and Zegada-Lizarazu and Monti [97] provide an overview of advantages of crop rotations:

— Reduced use of agrochemicals and synthetic fertilizers.

— Facilitate timing of farming activities.

— Lower erosion due to longer period of land cover.

— Improved soil structure.

— Improved soil texture.

— Improved soil fertility and higher yields.

— Maintenance of long-term productivity and organic matter.

— Improved population of microorganisms.

— Reduced number of weed seeds.

— Increased biodiversity.

— Greater market opportunities and lower economic and climatic risks due to diversified production.

An exhaustive discussion of these points is beyond the scope of this publication. Several agricultural publications describe them [84, 85, 95]. To provide a tangible example, we chose crop residues (e.g. straw and dead roots) that cause many interactions between crops grown on the same field in temporal succession. These effects are described in the following section.

2.3.1.3 Positive Crop-Rotation Effects: the Example of Crop Residues

Crop residues remaining on the field have great influence on the creation of positive crop-rotation effects. Even though the occurrence of crop residues is not restricted to crop rotations, as they may also occur in monoculture, they serve as good example for effects between different crops grown on the same field. The extent to which crop residues remain on the field depends on farming practices.

Table 29 Overview of crop residue effects to the soil; derived from Blanco-Canqui and Lal [94].

Type of properties	Effect	Explanation
Physical	Crop residues improve structural stability.	Protection of surface against erosive impacts of raindrops / water erosion and blowing wind / wind erosion.
Physical	Crop residues help avoiding surface sealing and crusting.	Surface sealing and crusting would negatively affecting hydrological properties in reducing water infiltration and runoff rate, inhibit seedling emergence, reduce air and heat fluxes and increase soil erosion
Physical	Crop residues improve the aggregate stability.	Aggregate stability helps to resist erosive forces, moderates freezing-thawing and wetting-drying cycles
Physical	Missing crop residues would lead to soil compaction.	Soil compaction would lead to clogging of macropores and reduction of pores connected to the surface, thus leading to less channels for earthworms and roots.
Physical	Crop residues improve hydraulic properties.	Hydraulic properties are e.g. total porosity, soil water retention and plant available water
Physical	Crop residues improve water infiltration into the soil and alter soil temperature dynamics.	This indirectly influences many other processes, taking place in the soil, e.g. seed germination, seedling emergence and growth, soil water storage, gaseous fluxes, microbial activities and nutrient availability.
Chemical	Crop residues affect presence of macro- and micro-nutrients.	The soil fertility and the yield are hereby affected.
Biological	Crop residues influence earthworm populations.	Crop residue-formed surface-connected macropores serve as nourishment source and habitat for earthworms; earthworms are essential to soil structural development, nutrient recycling, soil organic matter turnover, fluxes of water, air and heat across the entire soil profile.
Biological	Crop residues influence the dynamics of soil microorganisms.	Microbial activity stabilizes soil aggregates by producing organic binding agents.

We assume that the relevance of crop residues might be higher in deliberately created crop rotations, compared to rotations with less different crops in the rotation or systems without any crop rotation. Independently, whether in long crop rotations or in monoculture, crop residues remain on the field and affect the subsequent crops by positively influencing physical, chemical and biological soil properties, thus helping to maintain or even improve soil fertility from one crop to the subsequent one. In a review about crop residue impacts, Blanco-Canqui and Lal [94] describe functions of crop residues in the soil (**Table 29**).

The next section provides an overview of currently available methods to include crop-rotation effects in LCA.

2.3.1.4   Existing Approaches and Limitations for Inclusion of Crop-Rotation Effects in LCA

There is a consensus in agricultural LCA that crop residues, containing certain amounts of nutrients, should be considered in the assessment. Regarding phosphorous (P) and potassium (K), corrections can be performed for residues that either remain on the field – by allocating the respective environmental burden to the subsequent crop [98, 99] – or for residues that are removed from the field – by allocating the respective environmental burden to the harvested co-products [100, 101]. For nitrogen (N) remaining in crop residues on the field, a credit can be given if a reduced fertilizer dose is recommended for the subsequent crop [99].

Van Zeijts et al. [102] suggest allocating the burden associated with fertilizers providing N completely to one crop, allocating P and K according to the uptake and uptake efficiencies of the crops, and allocating the application of organic matter according to the land-use share of each crop in the rotation. Furthermore, they recommend to intensively study each agricultural activity in the rotation and to decide whether it is done for individual crops or more than one crop.

Positive effects in the crop rotation are not limited to N, P and K. Martínez-Blanco et al. [103] review several positive effects caused by compost as an organic fertilizer. The need to attribute these compost-related positive effects between the benefiting crops was underlined by Martínez-Blanco et al. [104], who described approaches for including them in LCA.

Here, N mineralization rates and N uptake were identified as the most promising approaches [105].

Besides the idea of including nutrient fluxes and several crops in LCA, it is worthwhile to think about the handling of the soil in the context of the system boundary. Cowell et al. [82] argue that the farmed soil should be within the system boundary, "because it is an integral part of the production system", and as the soil does not cross the spatial system boundary, soil quality must be taken into account. Similarly, Audsley et al. [106] add that soil crosses the temporal system boundary, and thus needs to be considered in LCA.

Alföldi et al. [107] included of a complete crop rotation in a LCA to compare organic farming versus conventional farming. This approach was deemed inappropriate for product-based LCAs, because the rotations were considered as a whole, and satisfactory allocation approaches down to the level of individual crops were not available.

These examples indicate that agricultural activities, e.g. fertilization and crop protection, are "meant to benefit more than one crop" [102]. If agricultural fertilization is studied, using a short observation period, e.g. one single vegetation period, there is a certain probability of overlooking several aspects of the overall fertilization strategy of the farm studied [108]. "This raises the question of whether it would be more appropriate to draw a system boundary around a crop rotation rather than a particular crop" [82].

As explained in the previous paragraphs, it seems inappropriate to consider only one vegetation period in agricultural LCAs, and it is recommended to consider broader time horizons. To provide a practical example and to convey an impression of temporal relationships, the amount of fertilizers to be applied and management practices, we refer to GRUDAF (Grundlagen die Düngung im Acker und Futterbau; Principles for fertilization in arable and fodder production). It was developed and released by the Swiss agricultural research institutes Agroscope Changins-Wädenswil ACW and Agroscope Reckenholz-Täntikon ART [108]. GRUDAF contains science-based recommendations for fertilization of arable crops and fodder. The document is oriented towards agricultural advisory services and farmers; it assists the development of economically and ecologically sound fertilization strategies. It is regularly updated [108]. The document reveals the complexity of fertilization and offers an excellent view into the vast number of factors influencing fertilization planning and fertilizer amounts. Based on scientifically validated long-term experiments, Flisch et al. [108] describe the following agricultural management aspects and soil properties that need to be considered to determine the fertilizer amount:

Management aspects:
— Crop rotation design.
— Types of previous crops.
— Usage of intermediate crops.
— Crop residue management.
— Number of grassland cuts or grazing of pasture.
— Long-term effects of organic fertilization (correction factors for second year after application).
— Animal-type-specific nutrient composition of organic fertilizer.
— Consideration of organic farming practices.
— Amount of precipitation during several time periods (e.g. outside vegetation period).

Soil properties:
— Mineralized N content.
— Soil organic matter content.
— Humus content.
— Clay content.
— Soil structure.
— Nutrient content.
— pH.
— Soil depth (shallow to deep).

For each of these factors and aspects, numerical correction factors are provided for adapting the actual fertilization practice [108]. Many of these aspects need to be considered at a time horizon broader than just one vegetation period. This reveals that the nutrient availability and uptake of individual crops are not only determined by fertilization activities starting with seedbed preparation. Instead, activities taking place months and even years before growing the considered crop significantly affect its quantity and quality [108]. Because this situation applies for all crops, one could state that the same error is acceptable for all of them — but one must acknowledge that each crop has different nutrition requirement profiles; thus, different situations may have different effects on different crops.

Besides removal of crop residues from the field and thus their associated nutrients, many further aspects contribute to crop-rotation effects. For example, changing the crop cultivated in a certain field helps to improve phytosanitary conditions. Also, the use of nutrients, water from different soil horizons and improvements in soil structure increases soil fertility and yields. Most of these examples of positive effects are plant-specific and have been supported by long-term field experiments.

Even though the previously described approaches are suitable for integrating the shift of nutrients from one crop to a subsequent crop in LCA, they do not seem to be widely used in agricultural LCA practice, and they fail to integrate the entire range of positive crop-rotation effects such as:

- Facilitated timing of farming activities,
- Improved phytosanitary conditions and reduced amounts of agrochemicals needed,
- Reduction of the probability of harvest failures and improved conditions for soil organisms,
- Improved soil texture, soil structure, root penetration and water availability,
- Improved soil fertility and increased yields.

These effects are not covered at all. Because of their relevance to agricultural practice, we suggest including these effects in the LCA methodology and propose a respective approach in Section 2.3.3. Before that, in the following section, we use the example of crop rotations to describe the need to include crop-rotation effects in LCA.

2.3.1.5   The Need to Include Crop-Rotation Effects in LCA: the Example of Crop Residues

A lack of crop residues on the field would lead to limited fulfillment of the aforementioned crop-residue effects. Soil fertility, crop growth and – over longer time frames – even the quantity of net primary production might be affected. Excessive removal of crop residue "adversely impacts sustainability of land use and cropping systems [, as well as] all the complex and dynamic factors influenced by residue removal [, ...] which also influence [...] soil compaction, [...] plant available water content, aeration, soil aggregation, soil tilth, SOM [soil organic matter] concentration, [...] nutrient storage and cycling" [94].

In times of limited (fossil and biogenic) resources, pressure is grow- ing towards intensification of crop-residue use outside of the field. At a first glance, crop residues seem to be easily available without negative consequences; thus, European laws are currently setting incentives for intensification of crop residue use. Examples include its use for heat production in straw-burning plants [68] and political incentives for additional use of crop residues as feedstock for the production of so-called 'advanced biofuels' [45, 109].

The European Renewable Energy Directive (RED) uses results of Carbon Footprints (CF) to score the 'environmental performance' of biofuels in order to verify their eligibility for political support. For this purpose RED defines a CF calculation method [45]. This method contains some weaknesses that may affect long-term soil fertility. In concrete terms, RED defines a new co-product group called 'residues'. By definition, these 'residues' are exempted from co-product allocation; thus, zero environmental impact from agricultural stage is allocated to them. In ISO-terminology, this treatment is reserved to 'waste'. In RED's softening of the ISO rules, we see a contradiction between the RED calculation method and established LCA methodology [21]. From agricultural and pedological points of view, the RED approach must be questioned because it establishes incentives for the removal of crop residues from agricultural fields, ignoring the existing functions of crop residues for soil quality and soil fertility and without setting any limits for their removal. Claiming environmentally friendly energy-provision and ignoring effects on soil fertility appears to be unbalanced.

The next section describes methods that are necessary for the proposed method to integrate crop-rotation effects into LCA.

2.3.2   Material and Methods

This section describes the method of mathematically describing crop rotations (Section 2.3.2.1), the general approach of expanding system boundaries and an agriculture-specific allocation approach (Section 2.3.2.2). A clear mathematical description of crop rotation helps LCA practitioners to keep an overview of the crop rotation in which the studied crop is grown, and in case of complex rotations, to derive probabilities that are needed later in the approach.

2.3.2.1   Mathematical Description of Crop Rotations

A systematic mathematical representation and classification of crop rotations was performed by Castellazzi et al. [110]. Types of crop rotations are fixed rotation, flexible cyclical rotation with fixed rotation length, flexible cyclical rotation with variable rotation length and flexible non-cyclical rotation with variable rotation length (**Figure 2**).

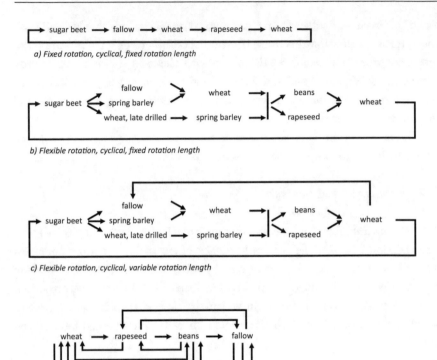

a) Fixed rotation, cyclical, fixed rotation length

b) Flexible rotation, cyclical, fixed rotation length

c) Flexible rotation, cyclical, variable rotation length

d) Flexible rotation, non-cyclical, variable rotation length

Figure 2 Examples of classified crop rotations; obtained from [110]

Crop rotations can be mathematically considered as a Markov chain — a mathematical model that describes transitions from one state to another and can be expressed by a stochastic matrix [111]. Stochastic matrices can be established for each type of crop rotation (**Table 30** and **Table 31**). The stochastic matrices provide probability of the occurrence of each crop in the rotation depending on the type of the previous crop [110].

One limitation of this mathematical representation is that only the crop grown in the year prior to the current crop is considered [110]; because Markov Chains "can determine the probability of any future state without reference to the past" [111], they characteristically do not have a memory. For the mathematical representation of crop rotations, this means that certain time intervals for growing one crop-type in the rotation (e.g. cultivation breaks due to phytosanitary restrictions) need to be carefully considered. Nevertheless, setting up correct stochastic matrices for crop rotations is possible by involving agricultural expertise when classifying the crop rotation (**Figure 2**) and deriving the respective stochastic matrices [110]. The availability of such mathematical representation is key for our method for considering crop-rotation effects in LCA. Besides a proper systematic mathematical description of crop rotations, an appropriate way is needed to allocate environmental interventions between the multiple products and co-products of the crop rotations.

Table 30 Stochastic matrix as mathematical representation of crop rotations – for examples a and b from **Figure 2**; derived from [110]

Crop rotation a		Current year				
		Sugar beet	Fallow	Wheat 1	Rapeseed	Wheat 2
Previous year	Sugar beet	0	1	0	0	0
	Fallow	0	0	1	0	0
	Wheat 1	0	0	0	1	0
	Rapeseed	0	0	0	0	1
	Wheat 2	1	0	0	0	0

Crop rotation b		Current year								
		Sugar beet	Fallow	Spring barley 1	Wheat, late drilled	Wheat 1	Spring barley 2	Beans	Rapeseed	Wheat 2
Previous year	Sugar beet	0	0.3	0.35	0.35	0	0	0	0	0
	Fallow	0	0	0	0	1	0	0	0	0
	Spring barley 1	0	0	0	0	1	0	0	0	0
	Wheat, late drilled	0	0	0	0	0	1	0	0	0
	Wheat 1	0	0	0	0	0	0	0.5	0.5	0
	Spring barley 2	0	0	0	0	0	0	0.5	0.5	0
	Beans	0	0	0	0	0	0	0	0	1
	Rapeseed	0	0	0	0	0	0	0	0	1
	Wheat 2	1	0	0	0	0	0	0	0	0

Reading note: All crops, occurring in the rotation are listed within the first line and the first column of the table. The first line represents crops currently grown; the first column represents crops previously grown. Within the cells of the table, one can read the probability for the current crop depending on the type of previous crop.

Reading example for crop rotation b: if sugar beet was the preceding crop in the previous year, with a probability of 35% late drilled wheat is to follow in the current year.

Table 31 Stochastic matrix as mathematical representation of crop rotations – for examples c and d from **Figure 2**; derived from [110]

Crop rotation c		Current year								
		Sugar beet	Fallow	Spring barley 1	Wheat, late drilled	Wheat 1	Spring barley 2	Beans	Rapeseed	Wheat 2
Previous year	Sugar beet	0	0.3	0.35	0.35	0	0	0	0	0
	Fallow	0	0	0	0	1	0	0	0	0
	Spring barley 1	0	0	0	0	1	0	0	0	0
	Wheat, late drilled	0	0	0	0	0	1	0	0	0
	Wheat 1	0	0	0	0	0	0	0.5	0.5	0
	Spring barley 2	0	0	0	0	0	0	0.5	0.5	0
	Beans	0	0	0	0	0	0	0	0	1
	Rapeseed	0	0	0	0	0	0	0	0	1
	Wheat 2	0.5	0	0	0	0	0	0	0	0.5

Crop rotation d		Current year			
		Wheat	Rapeseed	Beans	Fallow
Previous year	Wheat	0	0.2	0.5	0.3
	Rapeseed	0.5	0	0.25	0.25
	Beans	0.25	0.5	0	0.25
	Fallow	0.2	0.3	0.5	0

Reading note: All crops, occurring in the rotation are listed within the first line and the first column of the table. The first line represents crops currently grown; the first column represents crops previously grown. Within the cells of the table, one can read the probability for the current crop depending on the type of previous crop.

2.3.2.2   By-Product Allocation Approaches and Product-System Expansion

In addition to existing allocation approaches, such as mass allocation (based
on mass), energy allocation (based on lower heating value) and economic
allocation (based on market prices), a biophysical allocation approach based
on the Cereal Unit was proposed by Brankatschk and Finkbeiner [15]. The
Cereal Unit is a common denominator in agricultural statistics. It is suitable for vegetable and animal products [15]. Cereal-Unit conversion factors
are used to make all agricultural products and co-products comparable.
These factors are calculated mainly based on the products' nutritional values
for animals, taking into account the different digestive systems of various
livestock species; because 80% of the agricultural area in the world is used to
feed animals [15, 59].

The Cereal Unit allocation approach uses an agriculture-specific biophysical unit that was developed decades ago for the purpose of agricultural
statistics and is continuously updated and in use in German agricultural
statistics. The Cereal Unit is mainly based on the feeding value of each
agricultural product, and via auxiliary calculations, it is also able to represent products that are not directly fed to animals. By expressing all agricultural outputs of the inventory analysis in the same unit, using a common
denominator, all agricultural products become comparable and computable
[15]. This allows allocation of all agronomic inputs of the complete crop
rotation to each individual agricultural output regardless of whether it is of
vegetable or animal origin.

One alternative to avoid allocation is product-system expansion. This
approach is well known in the LCA community and recommended by ISO.
When applying system expansion, the product system is expanded "to include
the additional functions related to the co-products" [21].

2.3.3   Methodological Proposal

We describe a new approach to including the previously described crop-rotation effects into the life cycle inventory (LCI) of LCAs of agricultural products (according to ISO 14040 and ISO 14044). The steps presented are not meant to be an exhaustive description for performing an LCA. Rather, they should be understood as a supplement to the existing steps in the LCI analysis of an LCA. We would like to emphasize that the proposed modification of the system boundary is only relevant during the collection of data for the LCI and not the LCA as such. The overall scope of the LCA of an agricultural product/crop, its functional unit and its reference flow does not need to be changed.

In contrast to typical LCA, the proposed consideration of crop rotations expands the system boundary of the LCI, while not affecting the system boundary of the overall LCA study (**Figure 3**).

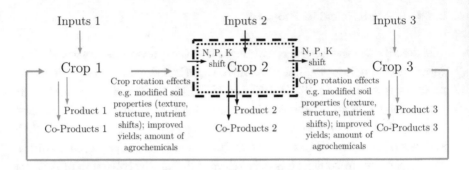

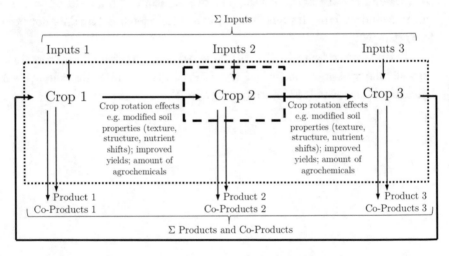

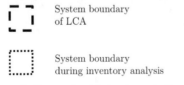
System boundary
of LCA

System boundary
during inventory analysis

Sum (Σ) of inputs are allocated between
all products, using their proportionate
share of the sum of all products, whilst
expressing them in the same unit –
calculation procedure described in
separate figure.

Figure 3 System boundaries in agricultural LCAs — comparison of current LCA practice
(upper part) and proposed method (lower part).

Figure 4 displays the structure of system boundaries within LCAs using the newly introduced approach, compared to current LCA practice. The proposed approach consists of the following steps:

1. The crop-rotation system is identified, in which the agricultural crop studied, e.g. wheat, is grown. The system boundary of the LCI (not that of the entire LCA study) is defined, including all elements around this crop-rotation system. The definitions of the functional unit and reference flow according to ISO 14040 (e.g. production of 1 t of wheat grain) thus remain unaffected. It is relevant at this stage to consider the entire rotation in which the wheat is grown.

2. The agronomic inputs (seed, diesel fuel, energy, agrochemicals, fertilizer, etc.) of the entire crop rotation cycle including all crops grown in the crop rotation are quantified.

3. All outputs (including products, by-products, waste, leachate, emissions) of this crop rotation leaving the agricultural field are considered and quantified (i.e. tonnages of each individual product, such as wheat grain and the other products and co-products produced within the same crop rotation).

4. All from agricultural outputs for each crop in the rotation are converted mass or volume into Cereal Units. Cereal-Unit conversion factors are available for Germany in various publications [15, 47, 49]. Cereal Unit conversion factors for other regions are not yet available but can be calculated without great effort [15].

5. Allocation factors are calculated for each individual agricultural output of the entire crop rotation using the amounts given in Cereal Units. Calculation check: the sum of all allocation factors must equal 100%.

6. Using the allocation factors calculated in the previous step, the sum of each agricultural input (seed, diesel fuel, energy, agrochemicals, fertilizer, etc.) is allocated among all individual agricultural outputs.

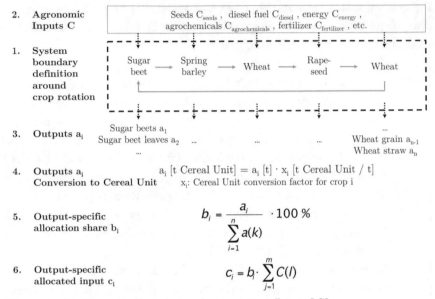

Figure 4 Calculation steps for inclusion of crop-rotation effects in LCI.

Thus, clear quantities of inputs and emissions are allocated to each individual output. The Life Cycle Impact Assessment (LCIA) uses the results calculated via this procedure. LCIA itself and later steps of LCA remain unaffected by this procedure. No further modifications are needed in the LCA.

For describing the new approach we used a relatively simple crop rotation (**Figure 4**). Moreover, the approach can be easily adapted to more complex crop rotations. Necessary for this is a mathematical description of the crop rotation in a stochastic matrix (**Table 30** and **Table 31**). Using these methods, every conceivable crop rotation can be mathematically described and probabilities for each crop within the crop rotation obtained. These probabilities are crop-specific and should be applied when quantifying all inputs and outputs of the crop rotation. Thus, the amount of a crop-specific input and the related amount of outputs are equally deduced from the probability that the crop occurs in the crop rotation.

Their contribution to the overall environmental burden and their individual share of the total outputs is thus crop-specifically calculated in accordance with their actual occurrence in the crop rotation.

A comparison between current LCA practice and the proposed method
was performed using a simple crop rotation. Winter wheat was assumed to
be the crop studied. For the sake of clarity, only two inputs (diesel as fuel
and N fertilizer) were selected for this comparison. Using current LCA
practice, 11.22 l of diesel and 30.87 kg of N fertilizer are needed to produce 1
t wheat grain (**Figure 5**). Applying the new method, 11.31 l of diesel and
25.57 kg of N fertilizer are needed to produce the same amount of wheat
grain (**Figure 6**). We would like to emphasize that results using the new
approach depend on the types of crops in the crop rotation. Different crops in
the rotation will lead to different results.

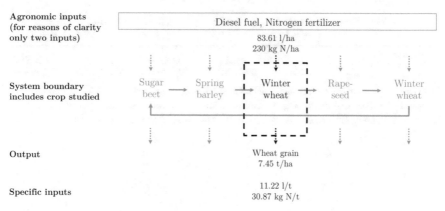

Figure 5 Calculation of specific inputs in current LCI practice — exemplarily for consumption
of diesel and nitrogen fertilizer in wheat production; data from [15, 57, 112-116].

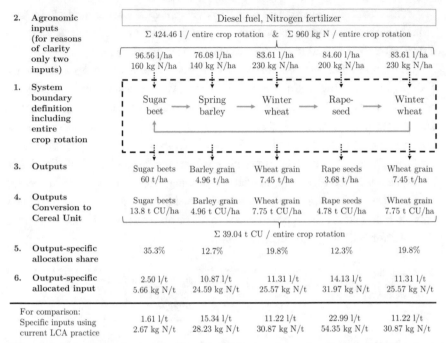

2.	Agronomic inputs (for reasons of clarity only two inputs)	Diesel fuel, Nitrogen fertilizer				
		Σ 424.46 l / entire crop rotation & Σ 960 kg N / entire crop rotation				
		96.56 l/ha 160 kg N/ha	76.08 l/ha 140 kg N/ha	83.61 l/ha 230 kg N/ha	84.60 l/ha 200 kg N/ha	83.61 l/ha 230 kg N/ha
1.	System boundary definition including entire crop rotation	Sugar beet	Spring barley	Winter wheat	Rape-seed	Winter wheat
3.	Outputs	Sugar beets 60 t/ha	Barley grain 4.96 t/ha	Wheat grain 7.45 t/ha	Rape seeds 3.68 t/ha	Wheat grain 7.45 t/ha
4.	Outputs Conversion to Cereal Unit	Sugar beets 13.8 t CU/ha	Barley grain 4.96 t CU/ha	Wheat grain 7.75 t CU/ha	Rape seeds 4.78 t CU/ha	Wheat grain 7.75 t CU/ha
		Σ 39.04 t CU / entire crop rotation				
5.	Output-specific allocation share	35.3%	12.7%	19.8%	12.3%	19.8%
6.	Output-specific allocated input	2.50 l/t 5.66 kg N/t	10.87 l/t 24.59 kg N/t	11.31 l/t 25.57 kg N/t	14.13 l/t 31.97 kg N/t	11.31 l/t 25.57 kg N/t
	For comparison: Specific inputs using current LCA practice	1.61 l/t 2.67 kg N/t	15.34 l/t 28.23 kg N/t	11.22 l/t 30.87 kg N/t	22.99 l/t 54.35 kg N/t	11.22 l/t 30.87 kg N/t

Figure 6 Calculation of output-specific allocated inputs using new method for inclusion of crop-rotation effects — exemplarily for consumption of diesel and nitrogen fertilizer in a crop rotation containing Sugarbeet, Spring barley, Winter wheat, Rapeseed, Winter wheat; data from [15, 57, 112-116].

2.3.4 Discussion

In current LCA practice, crop-rotation effects are only partly included, because it is difficult to quantify them, e.g. it is impractical to measure each nutrient flow in the soil — these data are not easily accessible, and they are not typically gathered in field experiments over decades. LCAs today typically assess each crop independently of its crop rotation and thus inadvertently ignore most of these crop-rotation effects, even though they are crucial for maintaining soil fertility and therefore are relevant for the sustainability of the agricultural system. To help consider these difficult-to-measure nutrient flows and difficult-to-quantify crop-sequence effects, we suggest this method. This supplemental approach is theoretically suitable for all agricultural LCAs. It takes into account all inputs and all outputs of the crop rotation and thus includes inter-crop relations, as well. For instance, this approach would make it possible to appreciate the benefit of legumes in fixing N and providing it to other crops in the rotation.

Within recent LCA practice, it is not obligatory to consider nutrient shifts from one crop to subsequent crops. Thus, fertilizing efforts are attributed to one single crop. This leads to free-rider situations for crops that consume nutrients left by preceding crops on the field (e.g. in crop residues). In this sense, subsequent crops are at an advantage and do not get charged for their true nutrient consumption because they receive some of the fertilization of the previous crop; subsequently, other crops within the crop rotation carry more environmental burden than is physically true. In other words, if LCA is performed for one individual crop, it does not consider that the crop may either enjoy the benefits or suffer the burden of being part of a crop rotation. In this context, it is worth mentioning that crops leaving high amounts of nutrients in the soil are, in recent LCAs without nutrient-compensating measures, systematically disadvantaged. This is because the full amount of fertilizers applied during their vegetation period is allocated to them, even if they do not consume all of it themselves, and some of their nutrients are transferred to the subsequent crop via their crop residues. By adapting system boundary to the crop-rotation level, free-rider phenomena for nutrient-receiving crops and systematic disadvantage of nutrient-lending crops can be avoided.

The free-rider phenomenon of advantaging certain crops has the potential to influence the conclusions drawn from LCA results. This situation is likely to occur when the preceding crop or crops are disregarded in the study; thus, the LCA practitioner is not able to depict the role of the preceding crop in the LCA result of the crop studied. For example, assuming growth of wheat after wheat versus wheat after rapeseed with identical inputs on the same agricultural field, the LCA practitioner will end up with higher environmental impacts for the former without understanding why different yields are achieved in the two scenarios, even though the same amounts of inputs were applied. This is due to the limited understanding of the role of preceding crops. Higher yields of wheat after rapeseed, and thus lower environmental impacts for the wheat compared to those of wheat after wheat, are caused by several reasons, such as improved soil texture due to interception of rainfall by rapeseed leaves in the previous year, improved soil texture due to root penetration of rapeseed plants and improved nutrient availability due to N transfer via crop residues. Another issue might arise if environmental impacts of a nutrient-lending and nutrient-receiving crop are compared to each other; here, the LCA results become flawed, decreasing the comparability of LCA results to the point that political decisions drawn from it have only limited justification. These examples indicate the need to include interactions among crops of a crop rotation in LCA.

Consideration of these positive effects using the proposed method is not restricted to the immediately subsequent crop. Improved conditions for subsequent crops throughout the entire crop rotation and over the course of several years are also taken into account by the modified system boundary. This is especially relevant for benefits that unfold over several years, e.g. phytosanitary effects, reduced need for agrochemicals and improved yields. Such effects are particularly difficult to measure directly or within a short time period; they could also easily exceed the budget of an LCA study. These effects can be integrated into the scope of LCA by including the entire crop rotation.

Taking the entire crop rotation into account is in line with consideration of other sectors in LCA: in the same way that each mineral oil- based
product is seen by LCA practitioners as a co-product of the refining process
of crude mineral oil and assessed or integrated in LCA databases as such, the
crop rotation is the overarching process for agronomists within which individual crops are grown. Each product can be seen as a co-product of the
entire crop rotation. While the process of crude mineral-oil refining seems to
be well-depicted in LCA, from an agronomist perspective, this is not the case
for crops as parts of crop rotations. The method proposed within this paper
allows practitioners to integrate the entire crop-rotation system into LCA.

In adjusting the system boundary to the crop-rotation level, one might
see a problem in different nutrient concentrations in the soil between the last
and the first crop of the crop rotation. But if the same type of crop rotation
is used on the same field in succession, the soil conditions at the beginning of
two identical crop rotations are the same. Therefore, these differences are
canceled out. This might not happen, however, if different crop rotations are
grown on the same field. For this situation, we suggest merging both rotations into one. This would not only resolve the previously described problem
but also increase the complexity of the LCA. This trade-off between result
accuracy and LCA complexity should be further explored in case studies.

The suggested alignment of the system boundary to the level of the crop
rotation brings LCA closer to reality, because it leads to an immediate
inclusion of all nutrient flows and changes in soil properties between crop-
rotation elements, e.g. through their effects on yields and on inputs needed,
without any need to actively measure them. Furthermore, the suggested
method offers new capabilities to LCA to treat crops more fairly, especially
those that leave nutrients on the field for the subsequent crop.

Existing approaches for including effects between crops are limited to
macronutrients. Current LCA methods are not suitable for expressing
whether the crop is grown in rotation or in monoculture. The LCA results of
a crop grown in a rotation are not identical to those of the same crop grown
in monoculture. Current LCA practice does not allow proper description of
these differences.

This approach changes the system boundary during the LCI. Consequently, all crops and thus the effects between them are included in the LCA. This happens because nutrient flows and crop-rotation effects between crops do not cross a system boundary without consideration. Both the subsequent crop and the studied crop belong to the same crop rotation; thus, the nutrients and positive effects remain inside the system boundary. Thus, the positive effects between crops are included via improved yields in the LCA.

Even though modifications to the system boundary are well-known in LCA practice, they are not widely used in LCAs for agricultural systems. One reason for that is the vast number of different outputs from one crop rotation and the enormous effort necessary to handle so many outputs. This is particularly the case when product-specific LCAs are performed. Integrating the entire crop rotation into the inventory analysis certainly introduces higher effort for practitioners, as well higher complexity to the assessment and – at a first glance – additional allocation problems are caused. But the LCA gains a better understanding of agricultural processes; the task of attributing environmental burdens to the vast number of different agricultural products and co- products can be shouldered by an agriculture-specific allocation approach, such as the Cereal Unit allocation. Of course, introducing allocation is connected to assumptions and value choices; an extensive discussion in the context of Cereal Unit allocation is given in Brankatschk and Finkbeiner [15]. The Cereal Unit is suitable for depicting all agricultural products and co-products. It organizes different agricultural outputs, making them comparable to each other and introducing computability. This serves as basis for an agricultural specific allocation approach [15]. Within the LCI, the Cereal-Unit-allocation approach is used to allocate all inputs uniformly to the individual outputs. Thus, agricultural inputs that are applied in one vegetation period but not used in the same period to grow crops or contribute to some extent towards the growth of other crops in the crop rotation can be more fairly attributed to the overall crop rotation. Thus, the crops currently regarded as 'single players', can be considered as 'team players' within the crop rotation, since their interactions are taken into account.

We would like to emphasize that the approach must not be confused with the already known and established system-expansion approach. Whereas system expansion affects the system boundary during all stages of LCA, our approach is clearly limited to adapting the system boundary at the level of the crop rotation when performing the LCI. Furthermore, an essential component of this new approach is an allocation step using the Cereal Unit allocation approach. Alternative agriculture- specific-allocation approaches might be suitable as well. This allocation step takes place as well within the LCIA. Here, the sum of all inputs of the crop rotation is allocated to their respective outputs using Cereal Units. Thus, no additional effort occurs for performing the next steps of LCA, and the functional unit and reference flow of the LCA remain unaffected.

Today, in many LCAs, environmental burdens are allocated mainly to the 'main' products, e.g. wheat grain. The relevant functions of co- products are poorly considered in some cases, e.g. the use of wheat straw as animal bedding in animal production, as fiber in animal nutrition or as a soil amendment to increase soil quality. Intentionally or unintentionally, co-products' functions are often disregarded. In the case of some political incentives, this might be an active choice, but it is unclear if side effects such as indirectly setting incentives that decrease soil quality were understood in the political decision process. From an LCA perspective, in the worst case, this could be partially caused by the lack of proper modeling of agricultural reality in LCA methodology. The present work shows a way to close this gap. Co-products' functions are inherently included within this approach, because allocating all inputs among all outputs is an elementary component of the method.

There is a logical relation between the amount of inputs, e.g. amount of fertilizers, that is needed to produce a certain quantity (yield) of agricultural outputs with a certain quality (composition). For some agricultural products, more input is needed to produce higher quality or quantity of outputs. A proper attribution of the inputs used between the agricultural outputs is needed, and account should be taken to quality of the individual outputs. This classical allocation problem can be solved by the Cereal-Unit-allocation approach for agricultural systems, because the Cereal Unit is mainly based on the nutritional value of agricultural products to animals. Thus, products with a greater nutritional value and greater benefit to animal nutrition get more inputs – and thus more environmental burden – assigned to them. In other words, the Cereal Unit is used to introduce the performance principle, based on animal nutritional value, to the attribution of environmental burdens within agricultural LCAs. The basis of that principle can be justified by the animal nutritional value as a biophysical principle, which is additionally encouraged by the fact that a vast majority of global agricultural area is used to feed animals [59], which serves as a basis for this approach.

2.3.5 Conclusions

In agricultural practice, the use of crop-rotation effects is an essential strategy. These effects are, for instance, based on improvements of nutrient availability, phytosanitary conditions and soil structure. This leads to increased yields and allows lower application rates of fertilizers and plant protection products. Against this background, crop-rotation effects are clearly relevant for assessing environmental impacts of agricultural crops.

Existing LCA practices have a limited ability to fully reflect these crop-rotation effects. Existing approaches are limited to particular effects only, such as shifts in individual nutrients. Limited consideration of crop-rotation effects within LCA was identified as a reason for the free-rider phenomena between the crops of crop rotations. These situations can affect overall LCA results.

To avoid such situations, a new method was proposed to supplement the established LCA methodology according to ISO 14040. This new crop-rotation approach adapts the system boundary to the level of the crop rotation for the LCI and uses an agriculture-specific allocation step to allocate inputs to their respective outputs. We suggest using the Cereal-Unit allocation approach, which represents animal feeding value. The new method allows practitioners to model the agri- cultural system over realistic time frames and includes fundamental agricultural consistencies such as crop-rotation effects in LCA methodology. The examples of crop-rotation effects given are well known in agricultural science.

The method integrates entire crop rotations in agricultural LCAs, including crop-sequence effects and establishing a performance-oriented attribution of environmental impacts between all agricultural outputs of the entire crop rotation. Positive crop effects are mirrored by im- proved yields in the entire crop rotation. The approach does not affect the functional unit and the Goal and Scope Definition. Thus, the method is suited to product-specific LCAs. The new approach helps LCA models to draw a more realistic picture of interactions between crops in a crop rotation and thus may help to further increase the reliability of LCA results. However, the approach should be tested using real-world case studies, and its results should be compared to LCA results performed using other recent methods. We encourage researchers to publish their results using this approach. These results would help practitioners understand whether the recommendations drawn from LCAs using this crop-rotation effect approach become more robust and whether the approach could help reach the target of sustainable development.

2.4 Crop Rotations and Crop Residues are Relevant for Carbon Footprints

This section contains the publication:
>   Brankatschk, G. & Finkbeiner, M. (2017). Crop rotations and crop residues are relevant parameters for agricultural carbon footprints. Agronomy for Sustainable Development, vol. 37, no. 6, p. 58; doi: http://dx.doi.org/10.1007/s13593-017-0464-4

>   The final publication is available at Springer via http://dx.doi.org/10.1007/s13593-017-0464-4

This paper illustrates qualitative impacts of using the Cereal Unit allocation approach (presented in section 2.1) and the crop rotation approach (presented in section 2.3) to the environmental performance of products. Each of the new methods is applied within case studies and the results are compared to those obtained using current modeling practice. Using reference studies from literature, the product carbon footprints (PCFs) of wheat bread, cow milk and rapeseed biodiesel are calculated using a one-year system boundary (current practice) versus applying the new crop rotation approach. Results show lower PCFs -11%, -22% and -16%, respectively. Hereby, numerical consequences of applying new methods are made visible (contributing to research target #9) and the product focus of the assessment is kept even though entire crop rotations were considered (#8). Furthermore, the PCF of wheat straw bioethanol was calculated applying the current modeling practice of allocating zero environmental burden to the straw versus allocating based on the Cereal Unit allocation, i.e. the animal nutritional value. PCF of wheat straw bioethanol increases by 80%, when allocating environmental burden between grain and straw (#2,9). Allocating zero environmental burden to agricultural residues, used for bioenergy purposes, is currently legally binding in the European Union. Without setting limits for straw removal from the field, this approach is identified setting incentives towards decreasing soil quality and not in line with the ISO standards for LCA. These findings underline the need for capturing long-term effects of agricultural practices (*#5*) and show the way towards ISO-conform modeling practice (*#9, 10*).

Abstract

Agriculture is the key for achieving the United Nations sustainable development goals: *food security* and *climate action*. To achieve these targets "climate-smart" agricultural practices need to be developed. Life cycle assessment and product carbon footprints are well established and inter- nationally recognized tools to assist the process of improving environmental performance. However, there is room for methodological improvement of agricultural life cycle assessments and product carbon footprints. For agronomists, it is widely known that crop rotations and crop residues do fulfill important agronomic functions, but they are not adequately represented in current life cycle assessment and product carbon footprint modeling practice. New methods tested in this study allow the inclusion of crop rotation effects and crop residues as co-products, whilst keeping at the same time the product focus. Product carbon footprints are calculated with and with- out consideration of these effects; results are compared. If crop rotations are considered, wheat bread, cow milk, and rapeseed biodiesel have lower product carbon footprints (− 11, − 22, and − 16%, respectively). The product carbon footprint of straw bioethanol significantly increases (+ 80%) when considering straw as an agricultural co-product instead of as waste. Ignoring crop rotation effects underestimates the annual greenhouse gas savings of EU-28 rapeseed biodiesel by 1.67 million t CO_2e and 20%, respectively. Here, we demonstrate for the first time that crop rotations and straw harvest should be considered for the product carbon footprints of bread, milk, and first- and second-generation biofuels. Since crop rotations and straw harvest are performed worldwide, the findings are relevant to all regions in the world. Comparing crop rotations and identifying climate-smart agricultural practices without losing the production orientation are key challenges for environmental assessments of agriculture in order to achieve the challenging combination of the food security and climate action sustainable development goals.

2.4.1   Introduction

Agriculture is the key to achieving the United Nations (UN) sustainable development goals (SDGs): *food security* and *climate action*. Population growth, climate change's impacts on agricultural yields, and reduced availability of arable land per capita (**Figure 7**) will lead to serious challenges in the coming decades. Besides being affected by climate change, agriculture itself has the potential to combat climate change [117-120].

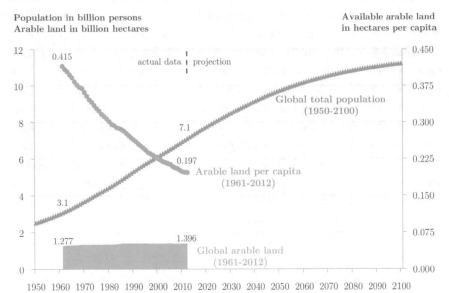

Figure 7 Amount of arable land per capita (1961–2012) is a function of global arable land (1961–2012) and global total population. Arable land per capita has approximately halved from 0.415 ha/person in 1961 to 0.197 ha/person in 2012.

The estimated increase of total global population (until 2100) will lead to a further decrease of arable land per capita. Thus, a further increase of agricultural productivity is needed to meet the future demand for food [121, 122]

There is a need to build highly resource-efficient agricultural systems, providing higher yields with less inputs, e.g., fertilizers [123]; to develop and assess regionalized farming tactics to increase food production with no cost to the environment [124]; and to perform sustainable intensification [125]. Life cycle assessment (LCA) and product carbon footprints (PCFs) are appropriate tools for accurately estimating environmental burdens of farming activities [3, 4].

PCFs estimate greenhouse gas emissions of products and have already gained legal relevance and market importance, such as carbon footprint labeling of consumer products. Since PCFs use LCA methodology, they have the same methodological strengths and weaknesses as LCA. Concerning agriculture, standard approaches of these methodologies fail to consider differences among agricultural management options [126], e.g., effects of crop rotations [127] and removing straw residues from fields (**Figure 8**)—two essential aspects of agricultural practice [16]. Fundamental challenges of integrating crop rotations and of co-product allocation in agricultural LCA are discussed in LCA community at least since the 1990s and since the 2000s, respectively. The magnitude of this gap has not yet been quantified.

Figure 8 Harvested and baled barley straw, ready for transportation

Here, we apply two new Life Cycle Inventory methods for the inclusion of these essential agricultural aspects: first is the cereal unit allocation approach for allocating environmental burdens among agricultural co-products and products. The basis is the biophysical cereal unit, which is based on the animal nutritional value. The cereal unit allocation is applicable to animal and plant products, traditionally used in German agricultural statistics and, in 2014, proposed as an agriculture- specific denominator for co-product allocation [15]. Second is the crop rotation approach, which combines system expansion and allocation within the Life Cycle Inventory in order to integrate effects among crop rotation elements. It is the first approach that allows considering crop rotation effects and, at the same time, maintaining the product focus of LCA [16]. Both of the methods are applied to show quantitative implications of excluding crop rotation effects and straw residues as co-products on the PCFs of wheat bread, cow milk, rapeseed biodiesel, and straw bioethanol.

2.4.2   Methods

For centuries, farmers have performed crop rotations to stabilize and improve yields [13]. In contrast, current product LCA and PCF modeling practices assess individual agricultural crops within a 1-year system boundary (option 1). This limits the ability to consider effects between crops grown in temporal succession on the same field. Approaches for modeling nutrient shifts between crops exist (option 2) but are limited to a few macronutrients and are not widely used [16]. We apply an approach for including crop rotation effects into LCA/PCF (option 3) and compare its results to option 1.

Since the 1990s, LCA practitioners have recommended including crop rotations and their effects on soil into LCA [106, 128]. These effects are not limited from one to the succeeding year. The following crop rotation effects are rather relevant on a longer time frame: "facilitated timing of farming activities, improved phytosanitary conditions and reduced amounts of agrochemicals needed, reduction of the probability of harvest failures and improved conditions for soil organisms, improved soil texture, soil structure, root penetration and water avail- ability, improved soil fertility and increased yields." [16]. To fill this gap, Brankatschk and Finkbeiner [16] extended the system boundary to the entire crop rotation and used an agriculture-specific allocation approach to allocate all inputs of the crop rotation among all outputs of the crop rotation. This system boundary extension and allocation of inputs is performed during the data collection step of each LCA, the Life Cycle Inventory (LCI) [16]. The following steps are used within this approach:

1. "The crop-rotation system is identified ... [and] the system boundary of the LCI (not that of the entire LCA study) is defined, including all elements around this crop-rotation system. The definitions of the functional unit and reference flow according to ISO 14040 (e.g. production of 1 t of wheat grain) thus remain unaffected ....

2. The agronomic inputs (seed, diesel fuel, energy, agrochemicals, fertilizer, etc.) of the entire crop rotation cycle including all crops grown in the crop rotation are quantified.

3. All outputs (including products, by-products, waste, leachate, emissions) of this crop rotation leaving the agricultural field are considered and quantified (i.e. tonnages of each individual product, such as wheat grain and the other products and co-products produced within the same crop rotation).

4. All agricultural outputs for each crop in the rotation are converted into Cereal Units ....

5. Allocation factors are calculated for each individual agricultural output of the entire crop rotation using the amounts given in Cereal Units ....

6. Using the allocation factors ..., the sum of each agricultural input (seed, diesel fuel, energy, agrochemicals, fertilizer, etc.) is allocated among all individual agricultural outputs." [16]

Since the 2000s, LCA practitioners have called for an agriculture-specific allocation approach, especially to overcome the problem of using different allocation approaches within the same agricultural system; doing so can over- or underestimate real burdens and is thus a source of uncertainty in agricultural LCAs [11, 34, 129]. To meet this demand, Brankatschk and Finkbeiner proposed the cereal unit allocation approach, which applies the established cereal unit from agricultural sciences to the methodology of life cycle assessment [15, 47]. The cereal unit is a common denominator for all (animal and plant) agricultural products and co-products; it is calculated based on the nutritional value of the products to animals and continuously updated and has been used in German agricultural statistics since the 1940s [15]. For the determination of a cereal unit (CU) conversion factor, the metabolizable energy content of different agricultural products is used and compared as a benchmark with the performance of barley. Therefore, 1 t of barley grains is equal to 1.00 t CU. As wheat has better animal nutritional parameters, 1 t of wheat grains is equal to 1.04 t CU. A detailed explanation of the calculation steps and a list of conversion factors for more than 200 agricultural products were published by Brankatschk and Finkbeiner [15].

Straw residues are agricultural co-products and influence soil structure, soil texture, and populations of soil organisms. They protect soil against erosive impacts of water and wind; they improve hydrological properties for infiltration and water runoff and enhance porosity, water retention, gaseous fluxes, heat fluxes, and availability of macro- and micronutrients; they provide habitat and nourishment source for soil organisms [16]. Thus, straw residues contribute to important soil quality parameters and consequently affect soil fertility and yields. As a consequence, harvested straw should be considered as a co-product in environmental assessments. Straw, remaining on the field, contributes to soil functions and does not cross system boundaries; therefore, remaining straw is not considered as a co-product. A differentiation is required between harvested straw and straw, remaining on the field.

The following procedure was performed to assess the influence of considering crop rotations and straw residues in results of PCF. First, existing

LCA or PCF studies were identified for wheat bread, cow milk, rapeseed biodiesel, and straw bioethanol. Since nitrogen fertilization is one of the largest contributors to agricultural greenhouse gas (GHG) emissions [130], it was used as the key variable. In the context of this study, the amount of nitrogen fertilizer was varied, whereas types of nitrogen fertilizers and emission factors remained unaffected. Other parameters (e.g., other fertilizers, processing, and transport) were fixed to exclude their influence on PCF results. Second, numerical contributions of nitrogen fertilization to the product's PCFs were identified and verified whether the PCFs did consider crop rotation effects. Third, the nitrogen fertilization, needed for the specific crop, was calculated using the previously explained crop rotation approach and the proportional deviation from the amount of nitrogen fertilization needed for the 1-year cropping system was calculated. Fourth, the GHG emissions related to the nitrogen fertilization of the 1-year system within the original PCFs were replaced by the respective GHG emissions for the system including crop rotations. Fifth, the PCF considering the crop rotation effects is obtained. This procedure aims at making differences between PCFs, using current methodology versus new methods visible. To some extent, this can be considered as sensitivity analysis for current modeling practice versus new methods.

2.4.2.1   Reference Studies

Several PCF studies exist for each of the selected products, but identification of the most accurate one for each product lay beyond the scope of this work. Selected PCFs should rather be understood as estimates and benchmarks for comparing current modeling practice to the proposed modeling approaches that consider crop rotations and crop residues.

Wheat bread has a PCF of approximately 460 g CO_2e/kg [131]. Approximately 57% of its GHG emissions relate to the agricultural stage [131] and 75% of them to nitrogen fertilization [130]. Therefore, 43% of its GHG emissions (approximately 200 g CO_2e/kg) are directly related to nitrogen fertilization.

Cow milk has a PCF of approximately 1240 g CO2e/L, of which 510 g CO2e/L is associated with feed production [132]. Assuming that 75% of the agricultural production relates to nitrogen fertilization of feed crops [130], approximately 380 g CO2e/L is directly related to nitrogen fertilization.

Rapeseed biodiesel has a PCF of 46 g CO2e/megajoule (MJ) [45]. Approximately 62% of its GHG emissions relate to the agricultural stage and 82% of them to nitrogen fertilization (11.0 g CO2e/MJ) and nitrous oxide emissions (12.5 g CO2e/MJ). The typical GHG reduction potential of rapeseed biodiesel is 45% compared to fossil diesel, based on the legally binding value of 83.8 g CO2e/MJ [45].

Straw bioethanol has a PCF of 11 g CO2e/MJ [45]. For its agricultural production phase, the EU Renewable Energy Directive (RED) states: "agricultural crop residues, including straw..., shall be considered to have zero life-cycle greenhouse gas emissions up to the process of collection of those materials" [45]. Thus, 0.00 g CO2e/MJ is used for the agricultural stage. The entire burden of wheat production is allocated to wheat grain, the main product. To calculate a PCF that includes agricultural emissions, the agricultural stage was modeled using the BioGrace calculation tool (see below); typical GHG emissions from the RED were added for processing (5 g CO2e/MJ) and transport (2 g CO2e/ MJ) [45].

2.4.2.2   Integrating Crop Rotation Effects and Crop Residues in PCF Results

GHG calculations and intermediate calculations for both biofuels are performed using the BioGrace tool, version 4d. This Microsoft® Excel-based calculation tool entails a harmonized GHG calculation methodology along the entire biofuel supply chain, including calculation of direct and indirect nitrous oxide emissions following the IPCC Tier 1 approach. BioGrace is recognized by the European Commission for calculating GHG emissions of biofuel production in compliance with the EU RED [133].

LCIs were generated for wheat (W), barley (B), rapeseed (R), and pea (P). Crops were chosen due to their relevance for European agriculture;

wheat and barley represent two thirds of the EU-28 cereal production, rapeseed is the main feedstock for biodiesel production, and pea was chosen as nitrogen-fixing plant. The crops are modeled both as individually grown (i.e., 1-year system boundary; see **Online Table 1**) and as elements of a crop rotation (R-W-P-W-B; **Online Table 2**). Mean yields and nutrient compositions of crops and crop residues were obtained from agricultural statistics and agricultural planning tables for Germany and other parts of Europe. In practice, crop rotations are often more complex than the chosen example. A mathematical representation of different crop rotation types helps considering complex rotations. Brankatschk and Finkbeiner [16] clarify calculation procedure for complex rotations.

Integration of crop rotation effects is tested for the PCFs of bread, milk, and biodiesel. For bread production, wheat grain is considered as an agricultural raw material. For milk production, barley grain, wheat grain, wheat straw, and rapeseed meal are considered inputs. For biodiesel production, rapeseeds are considered as a raw material. Consideration of crop rotation effects is performed using the previously mentioned method that takes place during Life Cycle Inventory only: firstly, extending the system boundary to the entire crop rotation, and secondly, allocating all inputs of the crop rotation among the outputs of the crop rotation using an agriculture-specific allocation approach—i.e., the cereal unit [16]. The resulting amount of nitrogen fertilizer per ton of each agricultural product entails the crop rotation effects. For the examples of bread, milk, biodiesel, and bioethanol, we assumed 1% of the straw to be harvested and 99% of the straw remaining on the field (Table 1). This amount of nitrogen fertilizer was used in the PCF calculation, resulting in PCFs of bread, milk, and biodiesel that consider crop rotation effects.

Attribution of environmental burdens to straw residues was tested for the PCF of straw bioethanol. For straw bioethanol production, wheat straw is considered an agricultural raw material. Whereas 1% straw harvest was assumed for the crop rotation part of the study, here, we assumed 100% of the straw being harvested and considered as a co-product. Using the cereal unit allocation approach [15], we calculated the amount of nitrogen fertilizer per ton of wheat straw (**Table 32**). This amount corresponds to the nitrogen demand used in the PCF calculation, resulting in a PCF that includes straw residues with environmental burdens.

Table 32 Overview of agricultural raw materials, considered for production of wheat bread, cow milk, rapeseed biodiesel, and straw bioethanol. Comparison of specific nitrogen inputs (kg N/t agricultural raw material) for the modeling options: "1-year systems" vs. "crop rotations" and "straw as waste" vs. "straw as co-product."
Compared to the current modeling practice, specific nitrogen inputs are lower when considering crop rotations and higher when considering straw as a co-product

Product	Nitrogen input for agricultural raw materials	
	1-year system	Crop rotation
Wheat bread	22.03 kg N/t wheat grain	16.70 kg N/t wheat grain
	19.94 kg N/t barley grain	16.06 kg N/t barley grain
	22.03 kg N/t wheat grain	16.70 kg N/t wheat grain
Cow milk	9.11 kg N/t wheat straw	6.91 kg N/t wheat straw
	1% straw harvest scenario	*1% straw harvest scenario*
	45.15 kg N/t rapeseeds	20.88 kg N/t rapeseeds
Rapeseed biodiesel	45.15 kg N/t rapeseeds	20.88 kg N/t rapeseeds
	Straw residues as waste (RED rules)	**Straw residues as co-product (ISO rules)**
Straw bioethanol	0.00 kg N/t wheat straw	6.87 kg N/t wheat straw
	100% straw harvest scenario	*100% straw harvest scenario*

Results are compared to PCFs based on current modeling practice, which ignores crop rotation effects and crop residues.

2.4.3   Results and Discussion

Remarkably different PCFs were found for wheat bread, cow milk, and rapeseed biodiesel, even though only nitrogen input was used as a variable. If crop rotations are considered, bread, milk, and rapeseed biodiesel have lower PCFs ($-$ 11, $-$ 22, and $-$ 16%, respectively) compared to current modeling practice (1-year systems) (**Figure 9**). If straw is considered as a co-product, straw bioethanol has a significantly higher PCF ($+$ 80%) than it does under the current modeling practice (**Figure 9**).

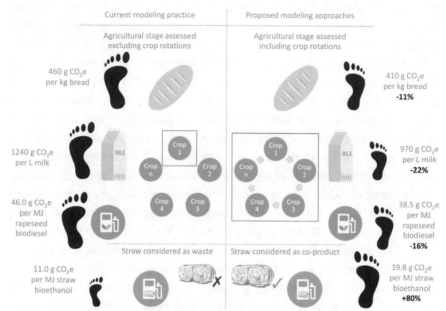

Figure 9 Comparison of product carbon footprints (PCFs) for wheat bread, cow milk, rapeseed biodiesel, and straw bioethanol using current modeling approaches: 1-year systems vs. crop rotation and straw as waste vs. straw as a co-product.
Considering crop rotations leads to lower PCFs (bread, − 11%; milk, − 22%; biodiesel, − 16%) and considering waste as a co-product to higher PCFs (straw ethanol, + 80%)

With an exception of nitrogen inputs, other parameters were fixed, in order to exclude their influence on PCF results. The differences presented mainly refer to the methods tested and to the aspects of crop rotation effects and crop residue allocation, which become measurable via the tested methods. Each method brings limitations. During application of the methods, some advantages and disadvantages were observed, which are shortly discussed below. An extensive method discussion would go beyond the scope of this paper. The advantage of the cereal unit allocation is the common denominator for assessing animal and vegetable products, which allows performing farm LCAs without changing the allocation approach. It helps avoiding unintended double counting or non-accounting of environmental interventions. Due to the physical relationship, based on animal feeding trials, the cereal unit allocation has a high ranking in the ISO hierarchy for allocation approaches. Whereas more than 200 cereal unit conversion factors do exist, they are only valid for German conditions. For use in other regions in the world, new conversion factors need to be calculated, which limits the applicability of the cereal unit allocation approach. The major advantage of the crop rotation approach is incorporation of crop rotation effects into LCA results whilst, at the same time, allowing product-based assessments. Temporal and spatial aspects of agricultural systems are taken into account. Hereby, product LCAs become able to represent whether the agricultural raw materials originate from improved crop rotations. Disadvantages of this method are additional data requirements for the entire crop rotation and additional workload for the LCA practitioner. Furthermore, the environmental interventions are attributed among all products according to the performance principle—when using the cereal unit as allocation, the animal nutritional value determines the allocation of environmental interventions. This also implies the attribution of interventions to all crop rotation elements, whereas these interventions may not occur for each of the elements in the crop rotation. For certain situations, e.g., for N fertilization of legumes, the combination of different ways of attributing bur- dens to crops may also serve as an interesting option; this aspect is explained in Goglio et al. [134].

The following subsections are focusing at considering crop rotations (2.4.3.1), considering crop residues (2.4.3.2), and sustainable agricultural practices (2.4.3.3).

2.4.3.1   Impacts of Considering Crop Rotations

Wheat grain assessed in 1-year system boundaries has a nitrogen input of 22.03 kg N/t (**Online Table 1**). When considering crop rotation effects, the nitrogen input is equal to 16.70 kg N/t wheat grain (**Table 32** and **Online Table 2**). Thus, the PCF of bread decreases from 460 g CO2e/kg by 50 g CO2e/kg to 410 g CO2e/kg, a decrease of 11% (**Table 32**, **Figure 9**).

Barley grain assessed in 1-year system boundaries has a nitrogen input of 19.94 kg N/t. When considering crop rotation effects, the nitrogen input is equal to 16.06 kg N/t barley grain, a decrease of 19% (**Table 32** and **Online Table 2**). Wheat straw assessed in 1-year system boundaries (and 1% straw harvest scenario) has a nitrogen input of 9.11 kg N/t (**Table 32** and **Online Table 1**). When considering crop rotation effects (also 1% straw harvest scenario), the nitrogen input is equal to 6.91 kg N/t wheat straw, a decrease of 25% (**Table 32**). Rapeseeds assessed in 1-year system boundaries have a nitrogen input of 45.15 kg N/t (**Online Table 1**). When considering crop rotation effects, the nitrogen input is equal to 20.88 kg N/t rapeseeds (**Online Table 2**), a decrease of 54% (**Table 32**). Based on these results, a reduction of 30% of nitrogen inputs for the feedstuffs has been applied. Thus, the PCF of milk decreases from 1240 g CO2e/L by 270 g CO2e/L to 970 g CO2e/L, a decrease of 22% (**Table 32**, **Figure 9**).

The 1-year system boundaries assessed for rapeseeds of 45.15 kg N/t nitrogen input (**Online Table 1**) show a good match with data from BioGrace of 44.14 kg N/t, which strictly follows the RED, and therefore represent European data [133]. Hence, BioGrace models individual crop growth within a 1-year system boundary; neither nitrogen transfer to subsequent crops nor crop rotations are considered. If crop rotation effects are considered, the nitrogen input is equal to 20.88 kg N/t rapeseeds (**Online Table 2**), a decrease of 54% (**Table 32**). The PCF of rapeseed biodiesel decreases from 46 g CO2e/MJ due to the reduced amount of nitrogen fertilizer by 5.8 g CO2e/MJ and due to the reduced direct and indirect nitrous oxide emissions by 1.7 to 38.5 g CO2e/MJ, a decrease of 16% (**Table 32**, **Figure 9**). In terms of GHG savings, the potential of 45% GHG reduction increases by 9 to 54% GHG reduction. Accordingly, the annual GHG savings of EU-28 rapeseed biodiesel consumption (approximately 6.0 million t rapeseed biodiesel) increases from 8.44 million t CO2e by 1.67 million t CO2e, or 20%, to 10.11 million t CO2e.

PCFs of bread, milk, and biodiesel are lower when including crop rotation effects, as some nitrogen remains as crop residues on the field and serves as fertilizer for subsequent crops. In contrast, within 1-year systems, total nitrogen demand is modeled as a fertilizer input, ignoring transfers of nitrogen between crops. Even though approaches for modeling nitrogen transfer from one crop to subsequent crops do exist, they apparently were not used in any of the PCF studies referenced [45, 130-133].

Interactions between crop rotation elements, such as nutrient flows, effects of soil organisms, and soil fertility, were identified as influencing PCF results. Moreover, crop rotations can improve soil nutrient resources and efficiency of nutrient use and even reduce the need for manure and chemical fertilizers [135]. These effects occur over longer time period than 1 year. Within the crop rotation planning, farmers do explicitly consider effects of individual crops. As team players con- tribute to the success of a team, individual crops contribute to the performance of the entire crop rotation. Current modeling practice does not distinguish between crops grown in rotations, in monoculture or in 1-year systems without nutrient transfer between crops. Thus, there is limited ability to compare environmental impacts of crops grown in different crop rotation systems; however, this ability is necessary to identify crop rotations with lower environmental impacts, to assist farmers in identifying "climate-smart" farming practices and to move towards both sustainable development goals: food security and climate action.

2.4.3.2 Impacts of Considering Crop Residues

Following the European legal definition for PCF calculation of biofuels, straw residues have a nitrogen input of 0.00 kg N/t wheat straw [45]; the GHG emissions of straw bioethanol are equal to 11 g CO_2e/MJ. We recalculate a PCF that considers straw residues as the co-product of wheat production. Here, we assume 100% of the harvest-ready straw being harvested (put by the combine harvester on a windrow, collected by a straw baler) and applied the cereal unit allocation that expresses the animal nutritional value [15]. Using the BioGrace tool, we calculate emissions of 1943.8 kg CO_2e/ha wheat and yields of 7640 kg wheat grain/ha, 6110 kg wheat straw/ha, and a nitrogen input of 168.84 kg N/ha. One kilogram wheat straw is used to produce 0.29 L straw bioethanol [136] or 6.173 MJ straw bioethanol/kg wheat straw; thus, 37,717 MJ straw bioethanol/ha is produced. Applying the cereal unit allocation to wheat grain (75.1%) and wheat straw (24.9%) [15], we calculate an environ- mental burden of 12.83 g CO2e/MJ straw bioethanol for the agricultural stage. After including the emissions of processing (5 g CO2e/MJ straw bioethanol) and transport (2 g CO2e/MJ straw bioethanol) [45], the total PCF is equal to 19.83 g CO2e/MJ straw bioethanol. Hence, the GHG emissions of straw bioethanol rise by 8.8 g CO2e/MJ, from 11 to 19.83 g CO2e/MJ, an increase of 80% (**Table 32**, **Figure 9**). In terms of GHG savings, the potential of 87% GHG reduction compared to fossil fuel decreases by 11 to 76% GHG reduction (see **Figure 10**).

The RED requires allocating zero environmental burden from the agricultural phase to straw [45]. According to ISO 14044, "it is necessary to identify the ratio between co-products and waste since the inputs and outputs shall be allocated to the co-products part only [21]. Hence, in an LCA and PCF context, the RED treats harvested straw as waste. This is not in line with the international LCA and PCF standards ISO 14040, 14044, and 14067 [16]. The RED definition means that the amount of straw necessary to produce a given amount of bioethanol is irrelevant to the latter's PCF. Consequently, there is no incentive to use resources efficiently. In addition, important functions of straw seem to be disregarded, such as those for animal bedding, animal nutrition, and improvement of soil fertility [16]. The overall relevance of crop residues to protection against erosion, nutrient recycling, carbon sequestration, humus balance, activity and diversity of soil organisms, holding capacity for nutrients and water, soil fertility, and thus future yields seems to be ignored [16, 137, 138].

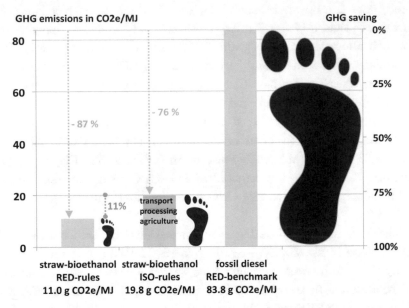

Figure 10   Comparison of product carbon footprints (PCF) of straw bioethanol: harvested straw considered as waste (calculation based on the EU Renewable Energy Directive; RED rules) vs. harvested straw considered as co-product (calculation based on ISO rules). The PCF of fossil diesel (obtained from the RED) serves as benchmark. The greenhouse gas (GHG) saving potential of straw bioethanol deviates 11 percentage points (87 vs. 76%)

Considering straw as waste in PCF calculations of biofuels could be politically motivated to promote non-food bioenergy feedstocks, avoid debates about food versus fuel, or provide an advantage to non-food-based biofuels when estimating reductions in GHG emissions. Another explanation could be a lack of understanding of the need to distinguish between harvested straw and remaining straw. Two systematic advantages are granted to straw-based biofuels. Firstly, they carry zero environmental burden from the agricultural stage, and secondly, their burden is allocated to their food-grade co-products. Food crop-based biofuels are hereby systematically disadvantaged. However, this approach seems unbalanced, because unlimited straw harvest affects soil fertility and, thus, future yields, including food yields on the same field.

Certain amounts of straw might be available for harvesting without affecting soil fertility [139]. These amounts depend on local conditions, may vary sharply, and should be defined in collaboration with soil scientists. A parallel line can be drawn to von Carlowitz's book about reforestation, *Silvicultura Oeconomica* [140]. It underlines the need for the balance between timber harvest and growth, formulating in this context the term *sustainability*. We contend that this principle should be transferred to straw harvest. Sustainable amounts of straw harvest should be defined that do not negatively affect long-term soil fertility and that ensure future yields. Furthermore, straw harvest practices may influence long-term soil carbon changes; increasing carbon in agricultural soils serves as carbon sink, whereas decreasing carbon content in soils leads to a release of carbon to the atmosphere. This aspect is not considered within this study.

2.4.3.3   Life Cycle Inventory Methods for Assessing Sustainable Agricultural Practices

Within this work, two new Life Cycle Inventory methods are applied and tested, i.e., the cereal unit allocation approach [15] and the crop rotation approach [16]. To ensure their compatibility with attributional LCAs, they should be conformed to existing international LCA standards, e.g., ISO 14040 and ISO 14044.

ISO 14044 provides a hierarchy for dealing with multi-output processes [21]: first, avoiding allocation via subdivision of processes in subprocesses or expanding the product system—which is hardly feasible for agricultural production processes that characteristically do have multiple outputs and would cause additional uncertainties [11, 15]; secondly, using physical allocation; and if not feasible, thirdly, applying economic allocation. The cereal unit allocation is based on animal nutritional value and therefore uses physical connections as basis for allocation [15]. The cereal unit allocation has been developed to overcome the co-product allocation challenge of agricultural production processes and serves as an agriculture-specific biophysical allocation approach. It is there- fore in line with the ISO standards for LCA.

Following the ISO hierarchy, higher priority should be given to biophysical allocation approaches compared to economic allocation. Aiming to include effects among crop rotation elements into attributional LCA, the crop rotation approach has been developed. It introduces a supplementary step into the existing LCI and does not affect other stages of LCA. It purposefully combines the well-known product system expansion (to the level of the entire crop rotation) with an agriculture-specific allocation approach. It can be applied with any allocation approach that serves as a common denominator for the agricultural products. In this study, the cereal unit has been used as the common denominator. Accordingly, the crop rotation approach is compatible to the ISO. Further details are provided in Brankatschk and Finkbeiner [16]. Therefore, both of the methods are in line with ISO 14044 and compatible to attributional life cycle assessments for agricultural systems.

Another important standard is the AGRIBALYSE method [141]. The AGRIBALYSE program started in 2009 and can be considered as a methodological standard in France. Herein, the French Environment and Energy Management Agency (ADEME) collaborates with a number of French and international organizations aiming to provide a consistent LCI database of French agricultural products. AGRIBALYSE databases are already publicly available and continuously updated. Background data are obtained from the ecoinvent database (version 3.1). The system boundary ends at farm gate, and therefore, primary processing of agricultural raw materials is excluded— e.g., grain milling and oilseed crushing. AGRIBALYSE defines an LCA assessment period of 1 year (harvest to harvest for plant production; January to December for livestock or permanent crops). When using the crop rotation approach, the same system boundary can apply at the LCA level, because the crop rotation approach only refers to the LCI. Hence, the tested crop rotation approach could assist future AGRIBALYSE versions in integrating crop rotation effects. With regard to the consideration of crop residues, i.e., straw, AGRIBALYSE suggests allocating the environmental burden among wheat grains and wheat straw using the economic allocation, acknowledging wheat straw is being harvested on 16% of the area and a wheat straw yield of 577 kg dry matter per ha. This is in line with the argumentation, provided in the previous section on impacts of crop residues and the need to allocate environmental burdens to harvested straw. But, the limited availability of straw prices prevents AGRIBALYSE from performing this allocation step in its current version [141]. The proposed cereal unit allocation as a biophysical allocation approach can overcome this problem of data availability and has even higher priority in the ISO allocation hierarchy. AGRIBALYSE already uses biophysical allocation approaches for livestock production. Using a biophysical allocation approach as well for plant products would be consistent within AGRIBALYSE. In that context, the cereal unit allocation can serve as a universal approach for both animal and plant production.

Comparison of methods tested within this paper and AGRIBALYSE method shows accordance with regard to the ISO standards. The concept of using biophysical relationships, which is the core of cereal unit allocation, is already partly integrated within AGRIBALYSE for livestock production and could be easily transferred to plant production, which even solves limitations in data availability and allows consistent co-product and crop residue allocation.

For centuries, crop rotations have been fundamental tools for securing and increasing yields [13]. To meet the challenges of future food provision and combating climate change, it is certain that crop rotations will gain additional relevance. There is urgent need to help farmers identify crop rotations that both reduce environmental impacts and respect the production function of agriculture. Reliable and accurate tools are needed to fulfill this task. Apart from considering nutrient carryover via crop residues [142], current modeling practice is not able to compare environmental burdens of different crop rotation options at the product level [16, 142, 143]. Usually, crops are individually modeled in a 1-year system boundary, even though the need for improvement has been recognized since the 1990s [16, 108, 128]. To make LCA and PCF methodology capable of supporting agricultural planning and drawing well-founded decisions towards sustainable agricultural practices [144], it is necessary to represent differences caused by different crop rotations [16].

The methods applied confirm a production-oriented performance principle in LCA and PCF. They use the agriculture-specific cereal unit, which is based on animal nutritional value [15]. It serves as a common denominator for all agricultural products and co-products and a variety of production systems. The animal nutritional value reflects future demand for food and feed better than the economic value or lower heating value used in the current LCA and PCF modeling [15]. By allocating environmental burdens to the same target, comparisons of different agricultural production systems become more reliable. In particular, comparing crop rotations and identifying climate-smart agricultural practices without losing the production orientation of agricultural systems are key challenges for environmental assessments in the next few decades—formulated in the sustainable development goals: *food security* and *climate action*. Similarly, potential impacts on soil fertility, and thus on future yields, of the crop residues left after harvesting should be given a closer look. This is in line with the overall trend in LCA development towards life cycle sustainability assessment (LCSA) that includes the environmental, economic, and social dimensions of sustainability [145-147].

2.4.4   Conclusion

This study demonstrates the influence of modeling practices and methodological weaknesses on environmental assessments of bioeconomy products such as food, feed, fiber, and biofuels. Crops were modeled either as 1-year systems or as crop rotations, and straw was treated either as waste or as a co-product. To date, these options have received little attention by PCF users or LCA practitioners. This study quantifies the impacts of different modeling options on PCF results. Hereby, sustainability scientists, political decision-makers, and the general public gain insights into challenges of model- ing agricultural production systems. In order to quantify the relevance of methodological choices and to derive information on the sensitivity of crop rotations and crop residues to LCA results, further case studies should be carried out and published.

The influence of potentially political decisions on PCF calculations, such as allocating zero environmental burden to straw used to produce bioethanol in Europe, is made visible. Consequently, environmental impacts of many agricultural products may be over- or underestimated. It is likely that public perception, political decisions, and even emission reporting of entire countries are affected. This study suggests that the ignoring of crop rotation effects leads to underestimation of annual GHG savings of rapeseed biodiesel in the EU-28 by 1.67 million t CO2e. For comparison, total biofuel GHG savings in Germany is equal to 5 million t CO2e. Because crop rotations are performed around the globe, the findings are relevant for environmental assessments of agriculture in every region of the world.

The tested modeling approach for crop rotations reveals as a real alternative to current modeling options and hereby supports the development and identification of sustainable agricultural practices. Without inclusion of crop rotation effects, environmental advantages of improvements in agricultural practices enabled by crop rotations would remain undetected. To keep pace with future needs and trends in agriculture and agricultural policies, crop rotations must be considered in LCA and PCFs.

We strongly recommend further testing and improvement of these methods, since they will be essential for evaluating the impacts of various agricultural management options. Reliable and meaningful assessment tools will be needed to help agriculture achieve the challenging combination of the SDGs *food security* and *climate action*.

2.5 Supplementary Material – Crop Rotations and Crop Residues are Relevant Parameters for Agricultural Carbon Footprints

This section contains material contains background data and calculations for the publication:

Brankatschk, G. & Finkbeiner, M. (2017). Crop rotations and crop residues are relevant parameters for agricultural carbon footprints. Agronomy for Sustainable Development, vol. 37, no. 6, p. 58; doi: http://dx.doi.org/10.1007/s13593-017-0464-4

This supplementary material comprises graphical representations of the underlying life cycle inventory (LCI) and its calculation behind the publication, presented in section 2.4. Furthermore, all background data, assumptions and calculation steps are provided. Provision of data and intermediate results provides transparency regarding changes of LCA results when using these new methods and enables immediate continuation of work in this field (*contributing to research targets #9,10*).

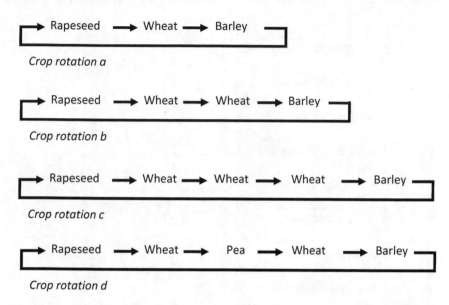

Crop rotation a

Crop rotation b

Crop rotation c

Crop rotation d

Figure 11 Crop rotations (*a*, *b*, *c* and *d*) including winter rapeseed, winter wheat, winter barley and spring pea

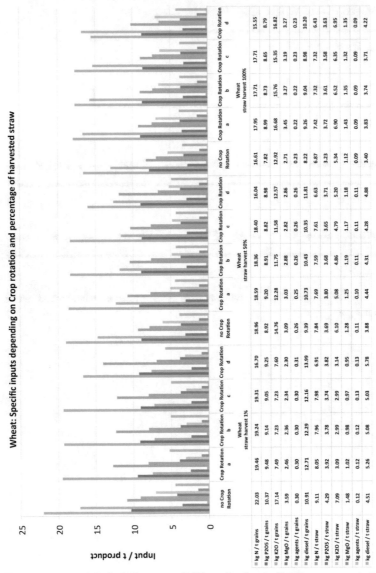

Figure 12   Life Cycle Inventory (LCI) results: Inputs (fertilizer, crop protection agents and diesel fuel) for wheat grain and wheat straw: influence of crop rotation and percentage of straw harvested

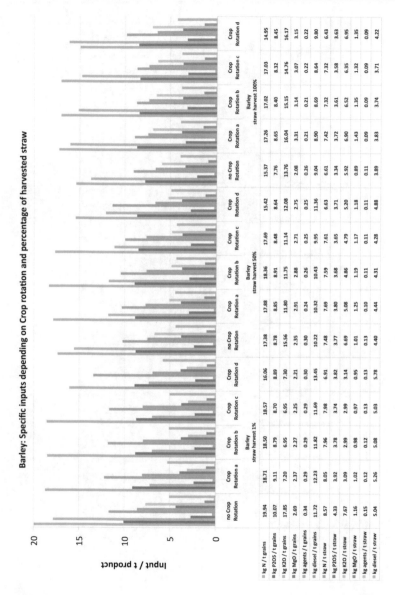

Figure 13 Life Cycle Inventory (LCI) results: Inputs (fertilizer, crop protection agents and diesel fuel) for barley grain and barley straw: influence of crop rotation and percentage of straw harvested

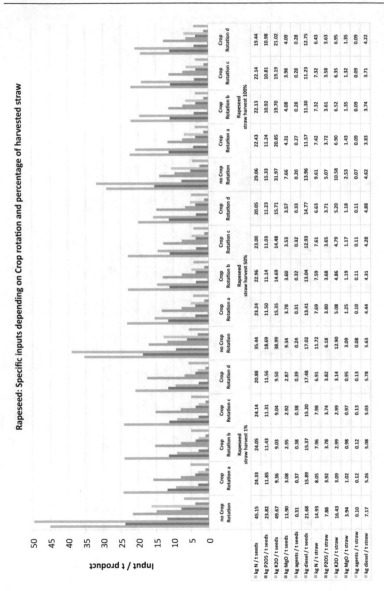

Rapeseed: Specific inputs depending on Crop rotation and percentage of harvested straw

Input / t product	Rapeseed straw harvest 1%					Rapeseed straw harvest 50%					Rapeseed straw harvest 100%				
	no Crop Rotation	Crop Rotation a	Crop Rotation b	Crop Rotation c	Crop Rotation d	no Crop Rotation	Crop Rotation a	Crop Rotation b	Crop Rotation c	Crop Rotation d	no Crop Rotation	Crop Rotation a	Crop Rotation b	Crop Rotation c	Crop Rotation d
kg N / t seeds	45.15	24.33	24.05	24.14	20.88	35.44	23.24	22.96	23.00	20.05	29.06	22.43	22.13	22.14	19.44
kg P2O5 / t seeds	23.82	11.85	11.43	11.31	11.56	18.69	11.50	11.14	11.03	11.23	15.33	11.24	10.92	10.81	10.98
kg K2O / t seeds	49.67	9.36	9.03	9.04	9.50	38.99	15.35	14.69	14.48	15.71	31.97	20.85	19.70	19.19	21.02
kg MgO / t seeds	11.90	3.08	2.95	2.92	2.87	9.34	3.78	3.60	3.53	3.57	7.66	4.31	4.08	3.98	4.09
kg agents / t seeds	0.31	0.37	0.38	0.38	0.39	0.24	0.31	0.32	0.32	0.33	0.20	0.27	0.28	0.28	0.28
kg diesel / t seeds	21.68	15.89	15.37	15.20	17.48	17.02	13.41	13.04	12.93	14.77	13.96	11.57	11.30	11.23	12.75
kg N / t straw	14.93	8.05	7.96	7.98	6.91	11.72	7.69	7.59	7.61	6.63	9.61	7.42	7.32	7.32	6.43
kg P2O5 / t straw	7.88	3.92	3.78	3.74	3.82	6.18	3.80	3.68	3.65	3.71	5.07	3.72	3.61	3.58	3.63
kg K2O / t straw	16.43	3.09	2.99	2.99	3.14	12.90	5.08	4.86	4.79	5.20	10.58	6.90	6.52	6.35	6.95
kg MgO / t straw	3.94	1.02	0.98	0.97	0.95	3.09	1.25	1.19	1.17	1.18	2.53	1.43	1.35	1.32	1.35
kg agents / t straw	0.10	0.12	0.12	0.13	0.13	0.08	0.10	0.11	0.11	0.11	0.07	0.09	0.09	0.09	0.09
kg diesel / t straw	7.17	5.26	5.08	5.03	5.78	5.63	4.44	4.31	4.28	4.88	4.62	3.83	3.74	3.71	4.22

Figure 14   Life Cycle Inventory (LCI) results: Inputs (fertilizer, crop protection agents and diesel fuel) for rapeseeds and rapeseed straw: influence of crop rotation and percentage of straw harvested

Life Cycle Inventory (LCI) data collection and results are presented in:

Online Table 3 Life Cycle Inventory (LCI) data collection for 1-year systems of Wheat, Rapeseed, Barley and Pea: Inputs (N, P2O5, K2O, MgO, Crop protection agents and diesel oil), outputs and conversion factors; Data obtained from [15, 55, 133, 148-157]

Online Table 4 Life Cycle Inventory (LCI) data calculation for 1-year systems of Wheat, Rapeseed, Barley and Pea: specific inputs per output (N, P2O5, K2O, MgO, Crop protection agents and diesel oil per ton of product or co-product); applying mass allocation, energy allocation, economic allocation and cereal unit allocation; Data obtained from [15, 55, 133, 148-157]

Online Table 5 Life Cycle Inventory (LCI) data calculation and results for crop rotation *a*; Data obtained from [15, 55, 133, 148, 150-157]

Online Table 6 Life Cycle Inventory (LCI) data calculation and results for crop rotation *b*; Data obtained from [15, 55, 133, 148-157]

Online Table 7 Life Cycle Inventory (LCI) data calculation and results for crop rotation *c*; Data obtained from [15, 55, 133, 148-157]

Online Table 8 Life Cycle Inventory (LCI) data calculation and results for crop rotation *d*; Data obtained from [15, 55, 133, 148-157]

Documentation of Data Sources

Yield data were obtained as average data 2008-2014 and 2006-2014 respectively from German Ministry of Agriculture [148] and Eurostat [149]. Nutrient concentrations of Nitrogen (N), Phosphorus pentoxide (P2O5), Potassium oxide (K2O) and Magnesium oxide (MgO) in agricultural crops and their co-products and Nitrogen-fixation rates of legumes were gathered

from planning tables of the Bavarian State Research Center for Agriculture
[150]. Consumptions of crop protection agents and diesel oil as fuel were
taken from Association for Technology and Structures in Agriculture [151],
BioGrace greenhouse gas calculation tool [133], The Thuringian State
Research Centre for Agriculture (TLL) [152, 153] and Eurostat [154]. Yield
increase data between crops in a rotation were taken from Richthofen et al.
[155] and Albrecht & Guddat [153]. Lower heating values, market values and
Cereal Unit conversion factors were taken from publications, market reports
of agricultural news websites [15, 55, 156, 157].

3 Discussion

Life cycle assessment (LCA) serves to identify environmentally sound options for production processes and for provision of services. In this context, LCA is used as decision-making tool. For the special case of agricultural production processes, several limitations in current LCA modeling practice were identified. Available modeling methodologies for agricultural LCAs are not capable to model agriculture to a level of detail that is needed to sufficiently represent agricultural management options and agricultural systems. This leads to a limited ability to derive robust recommendations towards farmers, consumers and political decision makers for climate-smart and environmentally sound farming practices.

This section is meant as overarching discussion. Therefore, it does not replicate all discussion strands of individual papers, presented in the previous chapter. For greater level of detail, please consider discussions presented in sections 2.1 (about co-product allocation approaches), 2.3 (about representing crop rotations in LCA) and 2.4 (about quantitative effects of considering crop residues and crop rotations towards LCA results). This section is structured in several sub-sections. Most of them are focusing at the method-oriented part of this thesis. The sub-section "Data, significant numbers and uncertainties" pays attention to the application-oriented part and the case study.

a) Co-product allocation in different frameworks, universal applicability of Cereal unit and the responsibility of LCA community
The current LCA practice of co-product allocation procedures is described causing non-accounting or double-counting of environmental interventions. These effects occur, when allocation procedures are prescribed in different frameworks, affecting the same agricultural system or primary processing steps, without realizing or considering the underlying connections. Examples for such frameworks, which are to some extend based on LCA methodology, are the European Renewable Energy Directive (RED) [45] and the Product Environment Footprint (PEF) Initiative of the European Commission [158].

© Springer Fachmedien Wiesbaden GmbH, part of Springer Nature 2019
G. Brankatschk, *Modeling Crop Rotations and Co-Products in Agricultural Life Cycle Assessments*, https://doi.org/10.1007/978-3-658-23588-8_3

In contrast to an LCA practitioner, who would only use one and the same allocation method for one processing step in one assessment, the RED and PEF will most likely define different allocation procedures for de facto the same processing step – leading to non-accounting or double-counting of environmental interventions (e.g. for oilseed processing: RED has defined energy allocation for vegetable oil, used for biodiesel production, and PEFs for oilmeals, intended for feeding animals or producing animal products, will use e.g. economic allocation; in this example a certain share of the environmental burden will not be accounted for; see section 2.1). This practice is in contrast to good LCA modeling practice, in conflict to ISO-series 14040 [15, 21, 22, 159] and reveals inadequate representation of agricultural supply chains.

Whilst RED and PEF are not strictly following LCA procedure, they are based on LCA methodology. For the LCA community, this situation creates a field of tension and certain responsibility. Whenever methodological application failures in LCA-based assessments appear, LCA community should point out these mistakes and provide advice to fix methodological challenges. The presented *Cereal Unit allocation* as agriculture-specific allocation approach can serve as solution to the described issue: whenever different modeling frameworks are using the same agriculture-specific allocation approach, the effects of non-accounting or double-counting of environmental burden can be avoided.

b) Vegetables, animals, residues – multifunctionality in co-product allocation
As common denominator for agricultural production including vegetable, animal and special agricultural products, the *Cereal Unit allocation approach* enables LCA practitioners assessing entire farm systems including animal and vegetable production – whilst at the same time allocating environmental burdens amongst crops, e.g. barley grains, and animal products, e.g. milk, more fairly, because the underlying procedure of allocating environmental burden between grains and milk is based on the same nutritional value. Exemplarily application of the new allocation approach to the allocation challenge milk-cow-calf (section 2.1) revealed that the *Cereal Unit allocation*

approach allows to include number of lactations and leads to allocation shares of 86.6% (milk), 6.8% (cow) and 6.6% (calf) – which are close to other biophysical allocation approaches [60, 62, 160], that were calculated based on greater level of detail.

As 80 percent of agricultural land is being used to feed animals [59] and thus representing its major utilization path, the Cereal Unit provides particular meaningfulness for assessments of agricultural systems. The Cereal Unit is based on the nutritional value and biological and physical connections. In expressing the nutritional value of co-products, it serves as rationale for allocating a certain share of environmental burdens to all agricultural co-products.

This can not be taken for granted – for instance the European Renewable Energy Directive RED requires to allocate zero environmental burden to harvested straw [45]. The political objective of promoting straw-based biofuels is prescribed in the carbon footprint calculation procedure for biofuels which is both legally relevant and a relevant aspect for the market [45]. Providing a zero-burden raw material to straw-based biofuel production might end up in very low greenhouse gas emissions, compared to fossil fuels. However, this procedure is in contrast to the current ISO standards for LCA [16], because criteria for being waste are not fulfilled. In addition, it might cause a shift of environmental interventions into other impact categories. Remaining straw on the field contributes to soil organic matter and humus balance. Therefore, certain amounts of straw are essential for maintaining soil quality and soil fertility.

Misleading incentives, such allocating zero environmental burden to the straw, without setting limits for its removal from the field, could have been avoided in using more substantiated allocation procedures. The reason to define allocating zero environmental burden to straw might be the limited ability to represent different qualities of agricultural products – e.g. when applying energy allocation to wheat grains and wheat straw, 55% of environmental burden would be allocated to grain and 45% to straw – which is hardly communicable or justifiable. Within this example, the energy allocation reveals on the one hand its robustness but on the other hand its impreciseness. Applying the *Cereal Unit allocation approach* – which is based on the animal nutritional value – the allocation shares are 75% to grain and 25% to straw and therefore serves as appropriate solution for this complex issue.

c) Geographical validity of conversion factors

Even though used since decades in official agricultural statistics in Germany, the Cereal Unit has been developed and updated for German conditions only. More precisely, this refers to the Cereal Unit conversion factors. These conversion factors are used to convert the various types of agricultural products and co-products into the Cereal Unit as common denominator. Therefore, the conversion factors are relevant for applying the *Cereal Unit allocation approach*. In a strict sense, the provided list of more than 200 Cereal Unit conversion factors is valid for Germany (and potentially Europe), which currently limits the geographical applicability of this approach. This limitation discouraged the authors of a study about energy demand and greenhouse gas emissions of different European breads in 2017 [161] applying the *Cereal Unit allocation approach*.

In order to apply the *Cereal Unit allocation approach* in other regions in the world – or even on a global scale – new regionalized lists of Cereal Unit conversion factors should be calculated. Required calculation procedure is described within this work (see section 2.2). Such calculation is out of the scope of this work and must include expertise from animal nutritionists. A new list of Cereal Unit conversion factors should undergo a critical review and the amount of work should not be underestimated [47]. Once Cereal Unit conversion factors are available for a studied region, the proposed *Cereal Unit allocation approach* itself can be applied without need for adaptation within every region of the world.

d) Conformity to ISO standards for LCA

According to ISO 14044, allocation should be avoided wherever possible. Due to multifunctionality of agricultural systems, allocation cannot be always avoided in LCA practice – especially for product LCAs (see section 2.1). When applying allocation, ISO states approaches based on "underlying physical relationships" shall be given higher priority compared to approaches without physical relationships, i.e. economic allocation [21]. Whilst mass allocation does not represent the inherent quality of a product – as it does not distinguish between 1 kg of water, 1 kg of protein or 1 kg of cellulose – and energy allocation expresses the value towards incineration, the *Cereal Unit allocation approach* allows comparing agricultural products, co-products and products from primary processing of agricultural products based on a common animal nutritional value. With regard to both of the presented methods, other LCA steps, i.e. goal and scope definition, life cycle impact assessment (LCIA) and interpretation remain unaffected. Consequently, the Cereal Unit allocation approach and crop rotation approach fit into the concept of ISO standardized environmental life cycle assessments.

Both the *Cereal Unit allocation approach* and the *crop rotation approach* are life cycle inventory (LCI) methods. Because the LCI is performed without limiting the type of impact categories, the presented methods are compatible to LCAs covering all conceivable impact categories. Whilst the case study presented within this thesis is focusing on product carbon footprints, the consequences of applying the new inventory methods are valid as well for other impact categories. Thus, the methods presented are compatible with existing impact categories and impact categories being developed in future.

e) Temporal and spatial aspects of agricultural practices
Farmers perform a broad range of agricultural practices with a temporal and a spatial dimension. For instance, they perform long-term fertilization strategies, which means fertilization of each nutrient does not take place every year and they aim for improved phytosanitary conditions, which is achieved by considering living conditions of plant pests. Furthermore, they deliberately distinguish the effects to the soil that are caused by different crop species, i.e. root penetration, nutrient leftover, nutrient use efficiencies, soil structure and soil tilth. Acknowledging these effects and further interdependencies, farmers perform the planning of cropping systems and crop rotations. This agricultural reality is in contrast to current LCA modeling practice, which has limited ability to distinguish whether an agricultural raw

material was produced in monoculture (no change of crop species over time) versus diverse crop rotations (different crop species on the same field over time). This example visualizes the relevance of temporal system boundaries for agricultural systems. Further limitations exist in the differentiation of spatial differences, e.g. several crops grown at the same time on the same field or crops grown in different regions or on soils with different fertility levels. This situation is highly unsatisfactory because improvements of agricultural practice are not reflected in LCA results. Various options of farmers' management choices cannot be modeled in LCA, which limits the capability of LCA as decision-supporting tool towards more sustainable, resource-efficient and climate-smart agricultural practices.

f) Crop rotation elements as team players
Having actual farming practices in mind, since 1990s, LCA practitioners have been admonishing for accounting interactions between preceding and subsequent crops [128]. From farmers' perspective farming systems consist of multiple years and the crops grown in temporal succession could be understood as "team players" (individual crops) in a "team" (crop rotation or cropping system). Each of the individual crops within a crop rotation contributes to a certain extend to the "success" of the entire crop rotation. Due to the nature of different crop species, different crops contribute and benefit to certain extend, e.g. via nutrient consumption, phytosanitary effects and effects to soil structure. These differences among crops lead to situations, in which a consideration of just one single crop would ignore that this crop is benefiting from other crops, without accounting for gaining that benefit. And vice versa the crop may provide benefits for other crops, whilst not accounting for such positive influence. These free-rider phenomena can be avoided by extending the system boundary to the level of the entire crop rotation in order to include all effects amongst crop rotation elements. Within the proposed *crop rotation approach*, a system expansion to the level of the entire crop rotation and an agriculture-specific allocation approach are combined. This procedure overcomes the challenges in temporal dimension and provides a solution for modeling entire crop rotations in LCA.

It should be noted that allocation of all inputs to all outputs of the rotation leads to situation that some impacts, relevant to one crop only, are shared amongst all crop rotation elements – for instance the seeding material. To some extend such situation already occurs whenever intermediate crops are grown, which are not harvested themselves and their burden is allocated to the following crop. Since each crop contributes to soil conditions, nutrient availability and phytosanitary effects, the allocation of these crop-specific burdens to all crop rotation elements could be considered negligible. The impact of these effects could be analyzed in further investigations, potentially leading to further methodological fine-tuning.

g) Product focus when assessing modified rotations
Consideration of the entire crop rotation into LCA by extending the system boundary to the level of the rotation needs additional data for creating the life cycle inventory, i.e. information about the rotation, inputs, yields, et cetera. Nevertheless, crop rotations have been already assessed before, e.g. in order to compare biological versus conventional agriculture. The novelty of the approach presented in this work, is the ability to maintain the product perspective and gaining ability to compare the product-related environmental burdens for products from different modified crop rotations. For example, the crop rotation approach allows LCA for wheat bread to distinguish between wheat grains, originating from different crop rotations. In other words, LCA methodology gains ability to measure differences in environmental performances for products, produced in different crop rotations. Crop rotations may be part of strategies towards climate-smart agricultural practices. This is relevant both, for distinguishing amongst monoculture and crop rotation, and for quantifying the environmental impact of integrating e.g. legumes (nitrogen fixing crops), fallow, intercrops, short- or long rotations. The allocation of environmental burden (of the entire rotation) to the individual outputs (of the rotation) enables to focus on individual products as part of a rotation. This is realized using the *Cereal Unit allocation* and would be possible with any other agriculture-specific allocation approach as well. In

absence of another universally applicable agriculture-specific allocation approach, the *Cereal Unit allocation* is the preferred option.

h) Challenge: simultaneously productive and climate-smart

As the *Cereal Unit allocation approach* is expressing the animal nutritional value of the individual products, this allocation approach is substantiated via biophysical interrelationship. Hereby, the proposed methodology entails the performance principle. This performance orientation allows LCA optimizing the quotient of environmental burden per production. Transferred to the UN Sustainable Development Goals (SDG) targets of *food provision* (SDG 1) and *climate action* (SDG 13), this methodology can be used to increase food provision, whilst at the same time decreasing climate impact of agricultural activity. Historically, the Cereal Unit has been developed as well in order to optimize the ratio between inputs and outputs of agricultural production. The use of the Cereal Unit within LCA transfers at the same time the concept of improving agricultural production and simultaneously keeping the environmental burden as low as possible. This may be used as a tool to find ways solving the challenging combination of SDG 1 *food provision* and SDG 13 *climate action*.

Current developments in LCA practice of relating environmental burdens per hectare will certainly lead to a decreased environmental burden per area (less activity equals less environmental burden), but, on the other hand, it can lead to a reduction of agricultural production and thus might insufficiently consider the production function of agriculture – this would be in contradiction to SDG 1 *food provision*.

i) Data, significant numbers and uncertainties

Data used within this work to calculate the life cycle inventories (LCI) and the product carbon footprints (PEF) have not been obtained from own field experiments. Therefore, these data are no primary data, but secondary data from literature. Data sources and literature references are described in detail in sections 2.2 and 2.5. As no primary data have been collected, an error analysis for this collection phase is not performed.

The focus of this work is not producing new LCA or PCF case study results. Instead, emphasis is given to developing and testing of new life cycle inventory methodologies versus established methods and to quantify consequences of using the new methods. Because results of testing new methods are better tangible with concrete numbers (e. g. PCFs), literature values have been used as reference. The selected literature does not raise the claim to be the most comprehensive or most accurate. Rather, the availability and transparence of intermediate results were relevant for using these studies as reference and basis for own calculations.

Contribution of this work is visualizing differences between LCI results or PCFs, calculated with current methodology, versus LCI results or PCFs, calculated using the new LCI methods, elaborated within this work. At first glance, it may appear irrelevant whether the reference value is 32 g CO2e/kg (i.e. 2 significant numbers) or 32.19 g CO2e/kg (i.e. 4 significant numbers). But in the context of applying and testing new methodologies, what matters are differences between the PCF results, obtained using current methodology versus those obtained using new methodologies. As a consequence, the number of significant numbers presented within this work should take into account that absolute numbers are being used for calculating relative differences between LCI results or PCFs. In order to calculate these relative differences without compromising accuracy, the absolute values should be expressed with a certain number of significant numbers. In this context, the count of significant numbers (or number of digits) must not be confused with expressing the level of accuracy for the absolute numbers itself. The count of significant numbers is rather needed as intermediate result for accurately and transparently calculating the relative difference between, e.g. PCF results, obtained applying current methodology versus new methods. This relative difference describes the quantitative consequence, when using a new methodology. **Table 33** illustrates the consequence of using 2 digits versus 4 digits.

Table 33 Numerical consequences of expressing product carbon footprints (PCFs) with 2 or 4 digits for the accuracy of relative differences amongst PCF results, obtained using different methodologies

	2 digits	**4 digits**
PCF 'current' methodology	32 g CO2e/kg	32.19 g CO2e/kg
PCF 'new' methodology	25 g CO2e/kg	24.56 g CO2e/kg
Relative difference between PCFs, obtained using current vs. new methodology	- 21.88 %	- 23.70 %

This example illustrates consequences of using 2 digits (- 21.88 %) and 4 digits (- 23.70 %) for comparing PCFs. It clearly demonstrates loss of accuracy when using 2 digits compared to 4 digits. This indicates the need for using 4 significant numbers. Calculations of this work were performed using Microsoft Excel without rounding intermediate results. Tables and diagrams, presented in this work, contain data with 4 significant numbers. Hereby, readers are enabled to follow all calculation steps and performing recalculations.

Calculation of Cereal Unit conversion factors is not part of this work, but they are substantial basis for this work. Therefore, it should be mentioned that the calculation of these conversion factors is exposed to certain imprecision. As the calculation of the conversion factors consists of several intermediate calculation steps, which are not performed within this work, it is hardly feasible to quantify errors for those steps. Nevertheless, a recommendation is made to calculate Cereal Unit conversion factors for further regions in the world. Therefore, a list of calculation steps is provided which individually contribute to an overall error:

— The calculation of gross energy for various agricultural products and co-products is performed using a standardized and established formula within animal nutritional sciences.

- The determination of animal specific metabolizable energy contents for various agricultural products and co-products is realized via animal feeding trials.
- The estimation of feed material shares, fed to specific animals, is estimated on the basis of statistics and expert information.
- Calculation of Cereal Unit conversion factors for animal products equals the amount of feed, expressed in metabolizable energy and is determined via animal feeding trials.
- Calculation of Cereal Unit conversion factors for vegetable products which are not used as livestock feed (e.g. strawberries) are calculated, taking into account three intensity levels of a reference crop. Classification into those three intensity levels considers economic production potential of the individual crop.

A detailed description of the calculation procedure is provided in section 2.2. As the error of the Cereal Unit conversion factor has not been quantified, no error bars are provided in graphical and numerical results of this work. In future calculations of Cereal Unit conversion factors, one should consider including an error analysis.

Despite this list of aspects contributing to a certain error within the calculation of Cereal Unit conversion factors, it is assumed that the quantitative effects of limited representation of spatial and temporal representation of agricultural systems and effects amongst crops grown on the same field are greater than the error introduced by the Cereal Unit itself.

j) Recent developments on allocation methods, crop rotations and use of presented methods

Besides allocation method presented within this work, further new allocation approaches for agricultural life cycle assessments were published. Selected examples are mentioned hereinafter. In 2015, a new co-product allocation approach was proposed, using the plant physiological construction cost of plant compounds [162]. This may serve as interesting approach for vegetable production systems and seems worth to be tested in practice. With regard to animal production systems, in 2017, a new allocation approach, based on the metabolic energy requirements of body-tissue growth [163], was proposed. This approach may assist solving co-product allocation problem of slaughter processes, where environmental burden needs to be allocated between various products and co-products. To some extend, both allocation approaches seem to be comparable and worth testing in future LCAs for agricultural and food processes.

The *Cereal Unit allocation approach* has been used in an Italian LCA for various straw management options in 2017 [164], was applied in 2016 in two Polish studies about Phosphorus efficiency [135] and Magnesium efficiency [165] of different crop rotations. In those studies, the Cereal Unit was used as common denominator to make outputs of different crop rotations comparable – this idea is compatible to the *crop rotation approach*. Furthermore, the *Cereal Unit allocation approach* was used in a 2016 Slovakian study for assessing the productivity of cereal and legume crops in rotations [166] and served as basis for the sensitivity analysis of an Italian mozzarella LCA case study in 2017 [167]. Another 2017 Italian study about conventional versus organic barley cultivation acknowledged described relevance of allocation amongst grains and straw and performed economic allocation between grains and straw [168].

Within a Danish farm study about barley cultivation including pig production in 2017 [169], the *Cereal Unit allocation approach* has not been used, because authors were missing conversion factors for co-products, e.g. pig meat and pig manure. This situation occurred, because the Cereal Unit only refers to living animals and does not serve as allocation for slaughtering processes. In future LCA studies, including animal production, one could reconsider using the functional unit of living animals – especially when animals are not slaughtered on farm and the system boundary ends at the farm gate and not at the plate of the consumer.

Within previously mentioned allocation approaches based on the plant construction cost, the body-tissue growth and the Cereal Unit, a tendency towards biophysical allocation approaches can be observed. These allocation approaches mainly refer to the ISO hierarchy for allocation approaches, which encourages LCA practitioners using allocation approaches based on physical relationships. Some LCA practitioners question this trend based on missing justification of causal relationship and just focusing of functional output [170]. Apparently, the debate about physical versus economic co-product allocation approaches is still ongoing. It can be foreseen, that the discussion about appropriate allocation approaches within the community of agricultural LCA is likely to continue.

In 2017 a review of 44 environmental assessment tools for agricultural products revealed that none of the tools is considering crop rotation effects [171]. The product carbon footprint (PCF) case study of this work (see section 2.4) for wheat bread, cow milk, rapeseed-biodiesel and straw-bioethanol using the *Cereal Unit allocation approach* and the *crop rotation approach* figured out differences between newly introduced modeling approaches and established modeling practice. The results of –11% (wheat bread); –22% (cow milk); –16% (rapeseed biodiesel) and +80% (straw-based bioethanol) emphasizes the importance of these modeling aspects and quantifies the need for a consideration of co-products and crop rotation effects. Further research groups have already started using the proposed *crop rotation approach*; e.g. for a multiple-cropping study of soybeans and sunflower in Brazil [172] and for assessing crop rotation design towards resource efficiency [173]. Within the Brazilian study the "Sunflower-soybean cropping system [was found to have] better environmental performance when compared to the combination of monocultures..." [172] Within a 2017 Swiss and Dutch long-term field experiment about the environmental impacts of cropping systems, tillage strategies and cover crops, the *crop rotation approach* including *Cereal Unit allocation* has been utilized: "To allow a comparison of our results with other studies, we expressed the final results per year by dividing the results by the duration of the complete crop rotation. In summary, we analysed the data per ha and year and per CU." [174]

In 2017, Peter et al. integrated to some extend crop rotation effects into a new assessment tool, called „Model for integrative Life Cycle Assessment in Agriculture (MiLA)" [175]. Here, PCFs and cumulative energy demands (CED) were calculated and compared to results of established modeling practice. The results show even bigger differences, compared to this work; i.e. PCF are varying –34% up to +99% and CED are varying –16% up to +89% [175]. These findings confirm calculations of this work and underscore to what extend consideration of crop rotation effects are actually influencing LCA and PCF results. The differences are too large to ignore them. Several authors announce considering crop rotation effects in future studies and acknowledge their relevance [176].

Another life cycle assessment tool for agricultural systems was published in 2017. This open source tool, called Crop.LCA, was developed in United Kingdom and Canada. It allows LCA screening of cropping systems, covers four categories (cumulative energy demand, global warming potential, acidification potential, eutrophication potential) and was tested for four cropping systems. The tool "can assess entire cropping systems over several years as a single entity" [177] and refers to the LCI methods presented within this work [15, 16].

In 2017, the discussion about integration of crop interactions achieved a new level in LCA community. Within a joint commentary and discussion article in the International Journal of Life Cycle Assessment, an international group of researchers (Goglio, Brankatschk, Knudsen, Williams and Nemecek) from Cranfield University (UK), Technische Universität Berlin (Germany), Aarhus University (Denmark) and Agroscope (Switzerland) [134] collected and arranged all available methods for assessing cropping systems. Each of the authors has already published about this topic and proposing individual approaches. The approaches were classified in three categories "allocation approaches", "Crop-by-crop approaches" and "Combined approaches". The category "allocation approaches" entails the presented cereal unit allocation and crop rotation approach. The group of authors intends to enhance visibility of considering cropping systems towards LCA practitioners and provides recommendations, depending on the objectives of the LCA study. It was clearly recommended to consider crop rotations and cropping system effects and to use a wise combination of the presented approaches. The cereal unit allocation and crop rotation approach are integral part of the recommended solution towards solving the challenge of considering crop rotation effects.

Aiming at supporting firstly, farmers in agricultural management decisions, secondly, legislators in political decisions and thirdly, consumers in their decisions, towards more sustainable, resource-efficient and climate-smart agriculture, assessment methods are needed that adequately represent agricultural systems and thus providing robust results and reliable recommendations. For improvement of life cycle based sustainability assessments, further investigation of current assessment practice of agricultural systems needs to be performed and further gaps being closed. With its contribution of an agriculture-specific allocation approach and the possibility of considering crop rotations in life cycle assessment, this work contains relevant contributions to life cycle inventory methods, helps to more realistically model agricultural systems and thus is an important step towards sustainable agriculture.

4 Conclusions

Within this work relevant challenges of agricultural LCAs were described, research questions formulated and ten research targets defined. These research targets were processed in a method-oriented part and an application-oriented part. Within the method-oriented part, two new life cycle inventory methods have been proposed. An illustration of numerical consequences of applying the new methods versus established modeling practice is provided in the application-oriented part. Each of the research targets was taken care of, as follows:

1. Consider multi-functionality of agricultural processes
The challenge of considering multi-functionality and multi-output in agricultural LCAs was described, discussed and the *Cereal Unit allocation approach* elaborated. This new allocation approach uses animal nutritional value as basis for allocating environmental interventions amongst all agricultural products and co-products. The use of Cereal Unit as common agricultural denominator introduces calculability amongst multitude of agricultural products. Hereby, multiple outputs and multi-functionality become manageable in agricultural LCA.

2. Avoid unintended double- or non-accounting of environmental burdens
Double-counting and non-accounting of environmental interventions may occur when using different allocation approaches in independent life cycle based assessments for (co-)products from one agricultural process in different sectors. As an agriculture-specific allocation approach, the *Cereal Unit allocation approach* is particularly suitable to be applied to agricultural systems. The use of a sector-specific allocation approach significantly reduces the probability of double- or non-accounting of environmental burdens, even though assessed in independent studies.

© Springer Fachmedien Wiesbaden GmbH, part of Springer Nature 2019
G. Brankatschk, *Modeling Crop Rotations and Co-Products in Agricultural Life Cycle Assessments*, https://doi.org/10.1007/978-3-658-23588-8_4

3. Consider vegetable and animal products in one allocation approach
The proposed *Cereal Unit allocation approach* is using a common denominator
for which conversion factors are available for both, animal and vegetable
products. Hence, this approach helps allocating burdens amongst animals and
vegetables and at the same time. It furthermore allows including the framework of entire farm systems, producing vegetable and animal products.

4. Common agriculture-specific denominator based on biophysical mechanisms
Whereas mass, energy and economic value do not reflect biological properties, several LCA research groups suggest using biophysical allocation
approaches. The *Cereal Unit allocation approach* is based on metabolizable
energy for animal nutrition and therefore on biophysical mechanisms. Side
calculations allow also considering agricultural products, not intended for
animal nutrition. Animal products are considered via the amount of feed,
needed for producing them. Therefore, the Cereal Unit is suitable as common
denominator for agricultural products and co-products. In reflecting animal
nutritional value, it represents the largest user of agricultural area in the
world.

*5. Mid- and long-term effects of agricultural management strategies and im-
proved phytosanitary conditions*
Multi-annual effects between crops and long-term agricultural farming
strategies, i.e. improved phytosanitary conditions, are identified being
relevant to agricultural LCA results. These effects are included in LCA via
using the proposed *crop rotation approach*. Within this life cycle inventory
(LCI) approach, temporal system boundaries are extended and burdens are
allocated amongst products. It allows integrating all effects between crops of
one rotation and allows considering the crop rotation elements as team
players. Free-rider phenomena amongst crop rotation elements are avoided,
but the amount of information about assessed agricultural systems increases.

6. Comparing environmental performances of different crop rotations
In order to provide farmers with recommendations towards environmentally
sound decisions, LCA methodology needs to reflect differences in environmental performance amongst different crop rotation strategies. The proposed
crop rotation approach allows performing LCAs for different crop rotations,
comparing their environmental performances and enables LCA methodology
for assisting farmers in improving sustainability of cropping systems.

7. Assess effects of integrating legumes, fallow and multiple cropping
Modifications of crop rotations, e.g. integration of legumes, enhanced crop
diversity, new crops, multiple crops per year, intermediate crops, cover crops
or fallow, end up in altered environmental performances. The *crop rotation
approach* allows modeling differences amongst modified crop rotations and to
compare environmental performances. This is urgently needed in order to
assist farmers in their planning with recommendations towards climate-
smart rotations.

8. Product-focus when assessing different crop rotations
Comparison of crop rotations should not end at the level of entire rotation;
rather, LCA methodology should be able to reflect the environmental burden
at product level. The proposed *crop rotation approach* purposefully combines
system expansion and biophysical allocation. The *crop rotation approach*
entails an attribution of the environmental intervention of the entire crop
rotation amongst all products and co-products of the rotation. This procedure for the first time allows keeping the product focus and allows comparing
environmental assessment results for products (e.g. wheat bread) that
originate from different crop rotations (e.g. from monoculture, short rotation, rotation with legume, rotation with intercropping, et cetera).

9. Numerical consequences of applying new approaches
For both of the new methods, the *Cereal Unit allocation approach* and the
crop rotation approach, numerical examples for applying the new methods
were provided on the life cycle inventory level. Furthermore, a product

carbon footprint case study was performed for wheat bread, cow milk, rapeseed-biodiesel and straw-based bioethanol. The results were compared to results obtained using current modeling practice. Differences are too large for ignoring them. This underlines the need to consider crop residues and crop rotations in LCA. Results of other research groups, already applying the new approaches, support this conclusion.

10. Ensure compatibility to standardized LCA methodology (ISO 14040 series) and availability of data for immediate use and further development
In providing application examples for the *Cereal Unit allocation approach*, for the crop rotation approach and in listing more than 200 Cereal Unit conversion factors, the methods can be further tested and immediately used for German conditions. Cereal Unit conversion factors for further regions should be calculated for use in other regions. Detailed explanations of necessary calculation steps and methodological background are provided. The *Cereal Unit allocation approach* is based on biophysical connections, which meets the criterion of physical relationship defined within the ISO standard for LCA. Both of the new methods are limited to the life cycle inventory only, which makes them compatible to various impact categories.

The Cereal Unit offers as common denominator solution for allocation discrepancies in modeling of agricultural systems and may help improving the reliability of LCA results of products and services derived from agricultural production. The *crop rotation approach* offers solution for verifying differences in crop rotation planning and comparing crop rotation performances on a product level. These approaches and the case study are contributing to achieve aforementioned research targets. This work entails contribution towards assessing temporal, special and multifunctional complexity of agricultural systems in LCA. In expressing the environmental burden of agricultural systems per common agriculture-specific unit, which at the same time expresses the nutritional value, allows to optimize the ratio between environmental burden per nutritional unit – this is a key to identify climate-smart agricultural production systems which at the same time contribute to the food production. In conclusion, the presented methods can be used achieving the challenging combination of Sustainable Development Goals *Food Security* and *Climate Action*. Further testing and methodological improvement are required for enhancing awareness within LCA-community and to establish crop rotation systems in LCAs. Research groups from several countries (Brazil, Canada, Danmark, Germany, Italy, The Netherlands, Poland, Slovakia, Switzerland, United Kingdom) already started to use the proposed methods. Parallel developments of biophysical allocation approaches, for considering cropping systems, crop rotations and crop interconnections should be compared to each other and bundled or combined. Further tasks for LCA community and prospects are described in the next section.

5 Prospects

Additional effort is required to further develop the methodology of agricultural LCA and to advance what has been achieved within this work.

From an application–oriented perspective, testing and applying the methods should be performed, which raises awareness within LCA community. In the coming years and decades, agricultural sector will need assistance for identifying climate-smart practices. Many questions will arise how to identify most sustainable option among various farming options. As crop rotations are of utmost importance for agricultural planning, consideration of crop rotations will be one of the core aspects for farmers. Compared to current situation, a much deeper link must be established between farmers and LCA practitioners. Both will need to better understand each other, which will improve the quality of results. Since method development takes certain time, LCA community should develop proper representation of agricultural production systems – including various options of farmers' management choices. This is required in order to prepare LCA for future demands of the agricultural sector. Agricultural management decisions towards climate-smart crop rotations should be based on robust assessment tools only. The proposed *crop rotation approach* is a contribution to the LCA community for being prepared for agricultural requests and helps to provide robust recommendations towards achieving more sustainable agriculture. The LCA community should test presented *crop rotation approach* and compare results to those, obtained with current modeling practice. This should include various different crop rotations, cropping systems and even LCA for entire farms. Different impact categories should be considered. This will create broader understanding of the relevance of crop rotation effects within LCA community and effects of rotations to specific impact categories. This testing will reveal further methodological limitations.

From a method-oriented perspective, further methodological improvements should be envisaged. Further steps should be taken to test, develop and establish the *Cereal Unit allocation* as agricultural allocation approach. Applying the method in practice will reveal further room for methodological

© Springer Fachmedien Wiesbaden GmbH, part of Springer Nature 2019
G. Brankatschk, *Modeling Crop Rotations and Co-Products in Agricultural Life Cycle Assessments*, https://doi.org/10.1007/978-3-658-23588-8_5

improvement and help identifying further research gaps. Missing Cereal Unit conversion factors should be calculated for further agricultural products, e.g. tropical products, fibre products (i.e. fibre plants, wood), agricultural co-products (i.e. species-specific straw), and further primary processing products (i.e. glycerin). Additionally, Cereal Unit conversion factors should be calculated for different regions or continents i.e. Africa, Asia, Australia, Europe, North America, South America. The results should be compared with existing list of Cereal Unit conversion factors for Germany. This work will allow understanding the representativeness of existing Cereal Unit conversion factors or the need for regionalized conversion tables. These steps lead the *Cereal Unit allocation approach* to a robust agricultural allocation approach, potentially applicable on country, continent or even global scale.

Whereas LCA has been traditionally focusing at environmental impacts, a new concept has been developed, which additionally includes economic and social dimension. This so called life cycle sustainability assessment (LCSA) entails economical, social, and environmental aspects [48]. As LCSA is based on LCA, LCSA contains the same methodological strengths and weaknesses. Whereas LCA is increasingly used to assess and improve environmental performance of products and services, obtained from agricultural production, one should also consider performing LCSA for agriculture. Since the methods proposed within this work focus on the life cycle inventory (LCI) and the LCI is largely consistent for LCA and for LCSA, the *Cereal Unit allocation approach* and *crop rotation approach* can be applied also within LCSA in order to improve the life cycle sustainability of agricultural production – which includes environmental, social and economic aspects.

Within its Sustainability Development Goals (SDG) [178] the United Nations describe key challenges of humankind to be solved in coming years and decades. The combination of SDG *Food Security* and SDG *Climate Action* creates an outstanding challenge for agriculture. Food security requires a significant increase in production, while emissions must be decreased. LCA can serve as important tool for assisting this process, however its methodology must acknowledge the nature of the contradicting objectives. The methods proposed within this work deliberately consider both, the need to decrease environmental burdens towards SDG *Climate Action* and the need to increase agricultural production towards SDG *Food Security*. Future work should consider these connections. A number of international initiatives deal with global challenges for agriculture. A corresponding overview is provided in the supplementary annex (chapter 6).

LCA is a valuable tool for assessing environmental impacts. Further methodological fine adjustments will help agricultural stakeholders for identifying climate-smart, resource-efficient and more sustainable agricultural management practices.

6 International Initiatives on Global Challenges for Agriculture

Whilst previous chapters are focusing at detailed methodological questions of LCA – and thus provide small mosaic pieces –, this chapter is meant to give an impression of the whole mosaic. This chapter provides a broad overview of global challenges for agriculture, derived from the sustainability development goals (SDG) of the United Nations. Furthermore, it contains an overview of international initiatives towards sustainable agriculture and the origin of the term sustainability. This section is not meant to be a political analysis – it is rather meant to describe the field of tension and steps already taken on a global level towards sustainable agriculture.

Relevance of land use and agriculture towards global sustainability is explained in sections 6.1 and 6.2. Approaches for achieving improved land use practices are briefly introduced in sections 6.3 and 6.4. Starting point of the term sustainability was 300 years ago; this is explained in section 6.5. The definition of sustainability was further refined over the last decades. Simultaneously, the need for reliable decision support described and, as described in section 6.6, life cycle based assessment of environmental effects was established.

6.1 Global Challenges and International Agreements

International, national and regional initiatives aim at facilitating the actions needed to fulfill ecological, economic and social targets. Accordingly, eight Millennium Development Goals (**MDG**) were defined in 2000 with regard to development, peace and collective security, human rights and the rule of law [179, 180]. "The MDGs helped to lift more than one billion people out of extreme poverty, to make inroads against hunger, to enable more girls to attend school than ever before and to protect our planet." [181] The Food and Agriculture Organization of the United Nations (FAO) states:

© Springer Fachmedien Wiesbaden GmbH, part of Springer Nature 2019
G. Brankatschk, *Modeling Crop Rotations and Co-Products in Agricultural Life Cycle Assessments*, https://doi.org/10.1007/978-3-658-23588-8_6

"since the early 1990s, the number of hungry people has declined by 216 million globally, a reduction of 21.4 percent, notwithstanding a 1.9 billion increase in the world's population... Despite overall progress, much remains to be done to eradicate hunger and achieve food security."[182]

Based on the MDG and learned sessions, new goals have been set for the period 2015 – 2030: the 17 Sustainable Development Goals (**SDG**) include all "three dimensions of sustainable development: the economic, social and environmental" [178]. *Goal 2* aims to "end hunger, achieve food security and improved nutrition and promote sustainable agriculture". [178] For this purpose, there is need to "double the agricultural productivity, ... ensure sustainable food production systems and implement resilient agricultural practices that increase productivity and production, that help maintain ecosystems, that strengthen capacity for adaptation to climate change, extreme weather, drought, flooding and other disasters and that progressively improve land and soil quality". [178] The intention of *Goal 13* is taking "urgent action to combat climate change and its impacts ... acknowledging that the United Nations Framework Convention on Climate Change [UNFCCC] is the primary international, intergovernmental forum for negotiating the global response to climate change". [178]

> *"2015 is a milestone year. We will complete the Millennium Development Goals. We are forging a bold vision for sustainable development, including a set of sustainable development goals. And we are aiming for a new, universal climate agreement."*
> *UN Secretary-General Ban Ki-moon* [181]

In its Paris meeting in 2015, the UNFCCC recognized "that climate change represents an urgent and potentially irreversible threat to human societies and the planet ... [and furthermore] ... fundamental priority of safeguarding food security and ending hunger, and the particular vulnerabilities of food production systems to the adverse impacts of climate change". [183] The UNFCCC Paris Agreement contains a clear target to hold the "increase in the global average temperature to well below **2 °C** above pre-

industrial levels and to pursue efforts to limit the temperature increase to
1.5 °C above pre-industrial levels". [183] Moreover, the agreement calls to
increase the ability to adapt to climate change, to promote climate resilience
and "low greenhouse gas emissions **development**, in a manner that **does
not threaten food production**". [183] Thus, the UNFCCC Paris Agreement establishes a link between climate change and food production.

6.2 Importance of Land Use Activities for Anthropogenic Climate Change

Besides being affected by climate change the agricultural sector itself significantly contributes to the global greenhouse gas emissions (GHG) and thus
for anthropogenic climate change. Based on information of the Intergovernmental Panel on Climate Change (IPCC), "the sector Agriculture, Forestry
and Other Land Use (**AFOLU**) **accounts for about a quarter** (~10 –
12 Gt CO2e / yr) **of net anthropogenic GHG emissions** [see
Figure 15] mainly from deforestation, agricultural emissions from soil and
nutrient management and livestock." [184]

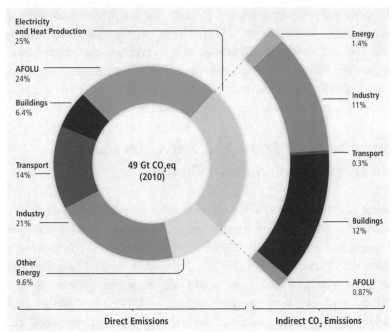

Figure 15   Greenhouse gas emissions by economic sectors; IPCC [184]

IPPC states, "most cost-effective mitigation options in forestry are af-
forestation, sustainable forest management and reducing deforestation, with
large differences in their relative importance across regions. In agriculture,
the most **cost-effective mitigation options are cropland management**,
grazing land management, and restoration of organic soils (medium evidence,
high agreement)." [184]

Within its Emission Gap Report (EGR) No. 6, the United Nations Environment Programme (UNEP) "provides a scientific assessment of the
impacts of the submitted Intended Nationally Determined Contributions
(INDCs) on anthropogenic emissions of greenhouse gases ... [The report]
compares the resulting emission level in 2030 with what science tells us is
required to be on track towards the agreed political target of a temperature
increase no more than 2°C by the end of the century. The report also pro-

vides data for the aspirational target of an increase below 1.5°C. In addition
the report analyzes selected areas where enhanced action can be taken and
how these actions can be accelerated and scaled up to close the 'gap'." [185,
186] Towards mitigation of greenhouse gas emissions, the UNEP and the
IPCC identify **agriculture** as "**key sector**" [185] and playing a "**central
role for food security and sustainable development.**" [185]

6.3   International Initiatives on Land Use

International initiatives have been established to assist reduction of emissions from agriculture, forestry and other land use (AFOLU).

6.3.1   Reducing Emissions from Deforestation and Forest Degradation
(UN-REDD / REDD+)

"The UN-REDD Programme is the United Nations Collaborative Programme on Reducing Emissions from Deforestation and forest Degradation
(REDD+) in developing countries. The programme was launched in 2008 and
builds on the convening role and technical expertise of the Food and Agriculture Organization of the United Nations (FAO), the United Nations Development Programme (UNDP) and the United Nations Environment Programme (UNEP). The UN-REDD Programme supports nationally led
REDD+ processes and promotes the informed and meaningful involvement of
all stakeholders, including indigenous peoples and other forest-dependent
communities, in national and international REDD+ implementation." [187]
"REDD+ is a mechanism that considers Reducing Emissions from Deforestation and Forest Degradation, including the role of conservation, sustainable management of forests and enhancement of forest carbon stocks in order
to create a financial value for the carbon stored in forests, offering incentives
for developing countries. UN REDD Programme supports national REDD+
readiness efforts in 60 partner countries." [186]

6.3.2   The New Vision for Agriculture by World Economics Forum

Against the background "greenhouse gas emissions and climate change increasingly threatens food systems" [188], partners of the World Economic Forum defined in 2009 The *New Vision for Agriculture*. It aims "to meet the world's needs sustainably **agriculture must simultaneously deliver food security, environmental sustainability and economic opportunity**. The Vision sets a goal of 20% improvement in each area per decade until 2050. Achieving those goals requires a transformation of the agriculture sector, leveraging market-based approaches through a coordinated effort by all stakeholders, including farmers, government, civil society and the private sector." [189] "The ... initiative engages over 500 organizations in its work to strengthen collaboration among relevant stakeholders. At a global level, it has partnered with the G7 and G20, **facilitating informal leadership dialogue and collaboration**. At the regional and country level, it has catalysed **multistakeholder partnerships** in 19 countries in Africa, Asia and Latin America, including **two regional partnerships called Grow Africa and Grow Asia**. Together, these efforts have mobilized over US $10.5 billion in investment commitments, of which US $1.9 billion has been implemented, reaching over 9.6 million smallholder farmers." [189] "The World Economic Forum's New Vision for Agriculture initiative supports national and regional partnership platforms engaging 18 countries across Africa, Asia and Latin America. ... Network members have co-created the New Vision for Agriculture Country Partnership Guide, a practical tool for practitioners to create and drive multistakeholder partnerships." [189]

6.4   FAO Activities Related to Agriculture and Climate

The Food and Agriculture Organization of the United Nations (FAO) established several frameworks, programmes and projects related to agriculture and climate. Some examples were selected and are briefly described in the following sections.

6.4.1   FAO-Framework: Climate-Smart Agriculture (CSA)

„Climate-smart agriculture (CSA) is an integrative approach **to address [the] interlinked challenges of food security and climate change**, that explicitly aims for three objectives:

(1) Sustainably increasing agricultural **productivity**, to support equitable increases in farm incomes, food security and development;

(2) Adapting and building **resilience** of agricultural and food security systems to climate change at multiple levels; and

(3) **Reducing** greenhouse gas **emissions** from agriculture (including crops, livestock and fisheries)." [190]

International, national and local actors rapidly adopted the term climate-smart agriculture. "However, implementing this approach is challenging, partly due to a lack of tools and experience. Climate-smart interventions are highly location-specific and knowledge-intensive. Considerable efforts are required to develop the knowledge and capacities to make CSA a reality. In large part, these are the same efforts required for achieving sustainable agricultural development which have been advocated over past decades, yet still insufficiently realized on the ground. CSA offers an opportunity to revitalize these efforts, overcome adoption barriers, while also adjusting them to the new realities of climate change. Organizations, educational establishments and other entities have started to fill these gaps, but **information is still fragmented**. A **partnership** between UN agencies (FAO, IFAD, UNEP, WB, WFP) and other organizations (CGIAR/CCAFS) has been created **to address knowledge gaps** and support countries in the implementation of climate-smart approaches." [119] The *Climate-Smart Agriculture – Sourcebook* was published to "... further elaborate the concept of CSA and demonstrate its potential, as well as limitations. It **aims to help decision makers** at a number of levels (including political administrators and natural resource managers) to understand the different options that are available for planning, policies and investments and the practices that are suitable for making different agricultural sectors, landscapes and food systems more climate-smart. This sourcebook is a reference tool for planners, practitioners and policy makers working in agriculture, forestry and fisheries at **national and subnational levels**. The sourcebook indicates some of the necessary ingredients required **to achieve a climate-smart approach to the agricultural sectors**, including existing options and barriers." [119]

"If current income and consumption growth trends continue, FAO estimates that **agricultural production will have to increase by 60 percent by 2050** to satisfy the expected demands for food and feed. ... Enhancing food security while contributing to mitigate climate change and preserving the natural resource base and vital ecosystem services requires the transition to agricultural production systems that are more productive, use inputs more efficiently, have less variability and greater stability in their outputs, and are more resilient to risks, shocks and long-term climate variability. ... By **reducing greenhouse gas emissions per unit** of land and/or agricultural product and increasing carbon sinks, these changes will contribute significantly to the mitigation of climate change. ... CSA is not a single specific agricultural technology or practice that can be universally applied. It is an approach that **requires site-specific assessments to identify suitable agricultural production technologies and practices**." [119]

"Examples of assessed benefits of **CSA practices** in **contributing to climate change adaptation and mitigation, and food security**:
   — Reduced soil erosion and improved nitrogen efficiency from minimum tillage, cover crops and **improved rotations**;
   — Improved water availability from soil and water conservation activities;
   — Improved crop yield with new varieties, a change in farm management (e.g. planting date change, fertilizer, irrigation water use) or short-term weather and climate forecasts;
   — Improved livestock productivity through enhanced breeding and feeding practice" [119]

Crop diversification "through **varied crop associations** and/or **rotations** (involving annual and/or perennial crops including trees)" [119] is of urgent relevance in Climate-smart agriculture.

6.4.2 FAO Framework: FAO-Adapt – Guidance for Climate Change Adaption

In 2011, FAO-Adapt was launched. The organization-wide framework "provides general guidance and introduces principles as well as priority themes, actions and implementation support to **FAO's multi-disciplinary activities** for climate change **adaptation.**" [190]

6.4.3 FAO Framework: Save and Grow – Sustainable Crop Production Intensification

"In 2011, FAO launched **Save and Grow** as a **new paradigm for intensive crop production** for that would **enhance both productivity and sustainability**. Save and Grow calls for greening of the Green Revolution through an ecosystem approach that draws on **nature's contribution to crop growth**, such as organic matter, water flow regulation, pollination and bio-control of insect pests and diseases." [190, 191]

"This eco-friendly farming often combines traditional knowledge with modern technologies that are adapted to the needs of small-scale producers. It also encourages the use of conservation agriculture, which **boosts yields while restoring soil health**. It **controls insect pests by protecting their natural enemies** rather than by spraying crops indiscriminately with pesticides. Through **judicious use of mineral fertilizer**, it avoids "collateral damage" to water quality. It uses **precision irrigation** to deliver the right amount of water when and where it is needed. The Save and Grow approach is **fully consistent with** the principles of **climate-smart agriculture** – it builds resilience to climate change and reduces greenhouse gas emissions through, for example, increased sequestration of carbon in soil.

For such a holistic approach to be adopted, environmental virtue alone is not enough: **farmers must see** tangible advantages in terms of **higher incomes, reduced costs and sustainable livelihoods**, as well as **compensation** for the **environmental benefits** they generate." [191]

6.4.4   FAO Framework: Global Plan of Action for Genetic Resources (GPA)

"Global Plans of Action seek to create an efficient system for the **conservation and sustainable use of genetic resources for food and agriculture**. They are intended as comprehensive frameworks to guide and catalyse action at community, national, regional and international levels through better cooperation, coordination and planning and by strengthening capacities." [190]

"The Second Global Plan of Action for Plant Genetic Resources for Food and Agriculture (Second GPA) is a strategic framework for the conservation and sustainable use of the plant genetic diversity on which food and agriculture depends. It was prepared under the aegis of the Commission on Genetic Resources for Food and Agriculture and adopted by FAO Council at its 143rd Session in November 2011.

The Second GPA reaffirms the commitment of governments to the promotion of plant **genetic resources as** an essen**tial component for food security** through sustainable agriculture in the face of climate change.

It is based on the findings of The Second Report on the State of the World's Plant Genetic Resources for Food and Agriculture and inputs from a series of regional consultations and experts worldwide. It updates the GPA on Conservation and Sustainable Use of Plant Genetic Resources for Food and Agriculture that was adopted by member countries in 1996. Updating the rolling GPA also strengthens its role in the implementation of the International Treaty on Plant Genetic Resources for Food and Agriculture. The Second GPA is thus current, forward looking and relevant to global, regional and national perspectives and priorities.

The Second GPA is an **agreed set of Priority Activities** that directly **address** the **new developments, opportunities and challenges** facing plant conservation and use in the 21st century. These include new policies and international agreements that affect conservation, use and exchange of genetic resources, shift in food production trends, changing roles of public and private sector in crop improvement and delivery systems, advances in biotechnology, genomics and information technologies, new products derived from agriculture, impact of new pests, climate change and rapid urbanization on plant genetic erosion and vulnerability. The priority activities of the Second GPA addresses these developments **to ensure** that **plant genetic resources for food and agriculture (PGRFA) continue to be available** for current and future use **for food security and sustainable agriculture.**" [192]

6.4.5   FAO Framework: Sustainable Land Management

Sustainable Land Management (SLM) was defined by the UN Earth Summit in 1992 as: "the use of land resources, including soils, water, animals and plants, for the production of goods to meet changing human needs, while simultaneously ensuring the long-term productive potential of these resources and the maintenance of their environmental functions".[190]

„[SLM] ... is crucial to minimizing land degradation, rehabilitating degraded areas and ensuring the optimal use of land resources for the benefit of present and future generations. ... SLM is considered an imperative for sustainable development and plays a key role in harmonizing the complementary, yet historically conflicting goals of **production and environment.** Thus one of the most important aspects of SLM is this critical merger of agriculture and environment through **twin objectives: i) maintaining long term productivity** of the ecosystem functions (land, water, biodiversity) and **ii) increasing productivity (quality, quantity and diversity)** of goods and services, and particularly safe and healthy food. To operationalize the sustained combination of these twin SLM objectives, it is essential to understand drivers and causes of land degradation and to take into account issues of current and emerging risks. " [193]

6.4.6   FAO Programme: Reducing Enteric Methane for Improving Food Security and Livelihoods

"The Emissions intensity (Ei) of **enteric methane** (CH_4) varies greatly across the globe. There are a number of ongoing efforts to **generate more robust estimates of mitigation potential in the livestock sector**. However, these efforts, are relatively new and fragmented and there is limited knowledge about the effectiveness and the applicability of mitigation measures over a range of regionally specific livestock production systems. In addition, there is a growing realisation that mitigation actions cannot be considered in isolation; true mitigation potential needs to consider 'packages' of actions assessed in terms of impacts on multiple gases and synergies or trade-offs between individual actions.

This project will **complement existing initiatives** to develop a complete picture of baseline emissions in beef production systems in South America (Argentina, Uruguay), and dairy production systems in East Africa (Ethiopia, Kenya, Tanzania and Uganda), West Africa (Benin, Burkina Faso, Senegal, Mali, and Niger) and South Asia (Bangladesh, Sri Lanka) while gathering information on already existing low-cost or no-cost mitigation measures, the barriers to uptake and the economic costs of using the measures. In order to **identify the most effective package of measures** that fit local farm systems, resources and capabilities, **and** to **avoid inadvertent trade-offs**." [190]

6.4.7   FAO Programme: Mitigation of Climate Change in Agriculture (MICCA)

"The Mitigation of Climate change in Agriculture (MICCA) programme contributes to global efforts to address climate change. It builds on the long-standing work carried out by FAO's technical departments and collaborates with international and national organizations. MICCA's work in developing capacities at local and national level, carrying out **pilot projects** and **generating technical knowledge** supports climate change actions **at the national level** as well as climate change negotiation processes undertaken through the UN Framework Convention on Climate Change (UNFCCC)." [190]

6.4.8   FAO Programme: Economics and Policy Innovations for Climate-Smart Agriculture (EPIC)

"The Economics and Policy Innovations for Climate-Smart Agriculture (**EPIC**) programme works with governments, research centres, universities and other institutional partners to support the transition to Climate-Smart Agriculture (CSA) by using sound economic and policy analysis. It is a **programme of work** aimed at **identifying and harmonizing** climate-smart agricultural **policies, analyzing impacts, effects, costs and benefits as well as incentives and barriers to the adoption of climate-smart agricultural practices.** The ultimate objective of the programme is to **support developing and in-transition countries** to formulate **agricultural investment proposals** to increase resilience to climate change and **promote CSA.**" [194]

"EPIC **analyzes** the **relative costs and benefits of changes in smallholder agricultural practices** in terms of CSA's three objectives: **adaptation, mitigation** and **food security.** The initial phase focuses on synthesizing available information on trade-offs and synergies globally. Next steps include costing prioritized CSA options resulting from data analysis and information available." [194]

6.4.9   FAO Programme: Integrating Agriculture in National Adaption Plans (NAPs)

"FAO and the United Nations Development Programme (UNDP) are joining forces to support countries as they integrate their agriculture sectors in the National Adaptation Plans (NAPs) process through the Integrating Agriculture in National Adaptation Plans Programme. The main goal is to support partner countries to identify and **integrate climate adaptation measures** for the agriculture sectors **into** relevant **national planning and budgeting processes**.

The Integrating Agriculture in National Adaptation Plans Programme initially targets eight countries: Kenya, Nepal, Philippines, Thailand, Uganda, Uruguay, Vietnam and Zambia. However, there are plans to expand the support to other countries in the Pacific, Asia, Africa as well as Latin America and the Caribbean. FAO support includes: - developing an integrated adaptation approach and roadmap

— Supporting key stakeholders in the process, in particular Ministries of Agriculture

— Defining a baseline on adaptation and identifying climate change knowledge gaps in the agriculture sector

— Developing and conducting capacity training

— Identifying climate finance for adaptation." [195]

6.4.10   FAO Programme: Livestock Environmental Assessment and Performance (LEAP) Partnership

In 2012, the Livestock Environmental Assessment and Performance (LEAP) Partnership was founded. It involves stakeholders "interested in improving the environmental performance of livestock supply chains. The programme's objective is to develop comp**rehensive guidance and a methodology for understanding the environmental performance of livestock supply chains**. The goal of this initiative is to contribute to the improved environmental performance of the livestock sector while **considering social and** economic viability. The Partnership provides **guidance on carrying out environmental assessments** and responding to the results. The Partnership promotes an exchange of data and information, technical expertise and research geared towards improving and **harmonizing the way in which livestock food chains are assessed and monitored.**" [196]

6.4.11   FAO-Project: Sustainability Assessment of Food and Agriculture Systems (SAFA)

"As the need for sustainable food and agriculture systems becomes increasingly urgent, there has been a variety of different sustainability initiatives launched in recent years with the aim of promoting a transition to sustainability. FAO built on existing efforts to develop this **universal framework for a Sustainability Assessment of Food and Agriculture (SAFA)**. After 5 years of participatory development, SAFA was presented to FAO member countries on 18 October 2013. SAFA can be considered like an **impact assessment tool** that is both **compatible and complementary** to most existing initiatives. The system is a holistic and inclusive framework for assessing sustainability performance in the food and agriculture sector, including crop and livestock production, forestry and fisheries." [197]

Regarding the relevance of crop rotations, SAFA acknowledges: "**Unacceptable conditions** and practices in relation to this objective: Crops are grown in monoculture, without any **crop rotation**, or only in a **two-year** constant **rotation with the same two crops**, although alternative crops would be available." [198]

6.5   Milestones in the Evolution of the Term 'Sustainability'

The term sustainability has an over 300 years lasting history. In his function as chief mining administrator of Saxony, Hanß Carl von Carlowitz realized a forthcoming scarcity of wood production. Wood was urgently required for mining activities. Carlowitz described the need for a continuous and sustaining balance between growth and harvest of wood. For this purpose, Carlowitz published in 1713 *'Sylvicultura Oeconomica, or the economic news and instructions for the natural growing of wild trees'*, a comprehensive book with practical instructions about reforestation. [140]

> *"Wird derhalben die größte Kunst/Wissenschaft/Fleiß und Einrichtung hiesiger Lande darinnen beruhen / wie eine sothane Conservation und Anbau des Holtzes anzustellen / daß es eine continuierliche beständige und nachhaltende Nutzung gebe / weiln es eine unentberliche Sache ist / ohne welche das Land in seinem Esse nicht bleiben mag."*
> *Carlowitz, 1713,* [140]

Further milestones, giving a meaning to the term 'sustainability', were in 1972 the UN Conference on the Human Environment in Stockholm, in1980 IUCN/UNEP/WWF/FAO/UNESCO World Conservation Strategy and the Brundtland Report 1987.

The Stockholm conference can be understood as the beginning of international environmental politics. "The environmental policies of all States should enhance and not adversely affect the present or future development potential ... and appropriate steps should be taken by States and international organizations..."[199] Even without mentioning the term 'sustainable development', it calls for a development, that considers economic, social and environmental dimension.

The World Conservation Strategy underlines the Earth as "only place in the universe known to sustain life" and therefore urges to conserve living and non-living resources as prerequisites for sustainable development. Three dimensions of sustainability and conservation were defined. Sustainable development "must take account of social and ecological factors, as well as economic ones ... [and] conservation is defined ... as the management of human use of the biosphere so that it may yield the greatest sustainable benefit to present generations while maintaining its potential to meet the needs and aspirations of future generations. Conservation, like development, is for people; while development aims to achieve human goals largely through use of the biosphere, conservation aims to achieve them by ensuring that such use can continue." [200]

"We have not inherited the earth from our parents, we
have borrowed it from our children." [200]

The Brundtland Report summarizes "sustainable development is development that meets the needs of the present without compromising the ability of future generations to meet their own needs." [201]

In 1988, the FAO council formulated its own definition with an orientation towards land use: "Sustainable development is the management and conservation of the natural resource base, and the orientation of technological and institutional in such a manner as to ensure the attainment and continued satisfaction of human needs for present and future generations. Such sustainable development (in the agriculture, forestry and fisheries sectors) conserves land, water, plant and animal genetic resources, is environmentally non-degrading, technically appropriate, economically viable and socially acceptable." [202]

During the UN Conference on Environment and Development (UNCED) in Rio de Janeiro 1992, a global partnership was created towards sustainable development. Important documents were adopted in order to achieve this goal. Hereunder, the Agenda 21, the Rio Declaration on Environment and Development, the Statement of Forest Principles, the United Nations Framework Convention on Climate Change and the United Nations Convention on Biological Diversity [203]. Supplemented by many further UN-conferences, the UNCED could be seen as key event, bringing sustainable development to the global agenda. Most recent meeting on the level of General Assembly of the United Nations was in 2015 the adoption of the 2030 Agenda for Sustainable Development including the Sustainable Development Goals (SDGs). [178]

6.6   Measuring Sustainability –
Life Cycle (Sustainability) Assessment
as Tool for Decision Support

The 1992 Rio Declaration on Environment and Development included the principle of 'environmental impact assessment'. It calls to "develop criteria and methodologies for the assessment of environmental impacts and resource requirements throughout the full life cycle of products and processes." [203] "Impacts within and among economic, social and environmental spheres" shall be included. [203] The "results of those assessments should be transformed into clear indicators in order to inform consumer and decision makers." [203]

In 1997, the international Standard ISO 14040 introduces such harmonized assessment. "The increased awareness of the importance of environmental protection, and the possible impacts associated with products [and services] ... has increased interest in the development of methods to better understand and address these impacts. One of the techniques being developed for this purpose is life cycle assessment (LCA)." [22]

In 2002 the United Nations Environment Programme (UNEP) joined its forces with Society of Environmental Toxicology and Chemistry (SETAC) to launch the **Life Cycle Initiative**, "an international partnership to put life cycle thinking into practice. The initiative is a response to the call from governments for a life cycle economy in the Malmö Declaration (2000). The mission of the Life Cycle Initiative is to develop and disseminate **practical tools** for the evaluation of opportunities, risks, and trade-offs associated with products and services over their entire life cycle to achieve sustainable development." [204]

During the World Summit on Sustainable Development (WSSD) in Johannesburg 2002, the world leaders also recognized that: "We must develop consumption and production policies to improve the products and services provided, while reducing environmental and health impacts, using, where appropriate, **science based approaches**, such as life cycle analysis".

According to the WSSD, life cycle approaches will have to play an essential role on the road towards **Sustainable Consumption and Production (SCP)**." [204]

"**Life cycle assessment (LCA)** is a tool for the **systematic evaluation of the environmental aspects** of a product or service system **through all stages of its life cycle**. It is standardised within the ISO 14040 series." [204] "An (Environmental) life cycle assessment (LCA) looks at potential impacts to the environment as a result of the extraction of resources, transportation, production, use, recycling and discarding of products; life cycle costing (LCC) is used to assess the cost implications of this life cycle; and social life cycle assessment (S-LCA) examines the social consequences." [78] "However, in order to get the 'whole picture', it is vital to extend current life cycle thinking to encompass **all three pillars of sustainability: (i) environmental, (ii) economic and (iii) social**. This means carrying out an assessment based on environmental, economic and social issues – by conducting an overarching **life cycle sustainability assessment (LCSA)**." [205]

Thus, life cycle assessments are appropriate tools for measuring sustainability and supporting informed choices on environmental, economic and social sound options. "The responsibility of the researchers involved in the [life cycle] assessment is to provide appropriate and reliable instruments." [206]

References

[1]   M. Finkbeiner, A. Inaba, R. B. H. Tan, K. Christiansen, and H. J. Klüppel, "The new international standards for life cycle assessment: ISO 14040 and ISO 14044," (in English), International Journal of Life Cycle Assessment, Editorial Material vol. 11, no. 2, pp. 80-85, Mar 2006.

[2]   N. Holden, "10th International Conference on Life Cycle Assessment of Food 2016," in 10th International Conference on Life Cycle Assessment of Food 2016, Dublin, Ireland, 2016, p. 1431: University College Dublin.

[3]   R. Schenck and D. Huizenga, "Proceedings of the 9th International Conference on Life Cycle Assessment in the Agri-Food Sector (LCA Food 2014)," in 9th International Conference on Life Cycle Assessment in the Agri-Food Sector (LCA Food 2014), San Francisco, 2014, p. 1574: American Center for Life Cycle Assessment, 2014.

[4]   H. M. G. van der Werf, M. Corson, and A. Wilfart, "LCA Food 2012—towards sustainable food systems," (in English), The International Journal of Life Cycle Assessment, pp. 1-4, 2013/03/28 2013.

[5]   M. S. Corson and H. M. G. van der Werf, "Proceedings of the 8th International Conference in Life Cycle Assessment in the Agri-Food Sector (LCA Food 2012)," in 8th International Conference in Life Cycle Assessment in the Agri-Food Sector (LCA Food 2012), Saint-Malo, France, 2012, p. 955: INRA, Rennes, France.

[6]   B. Notarnicola, "7th International Conference on Life Cycle Assessment in the Agri-Food Sector (LCA Food 2010), 22–24 September 2010, Bari (Italy)," (in English), The International Journal of Life Cycle Assessment, vol. 16, no. 2, pp. 102-105, 2011/02/01 2011.

[7]   G. Gaillard and T. Nemecek, "6th International Conference on LCA in the Agri-Food Sector," (in English), The International Journal of Life Cycle Assessment, vol. 14, no. 7, pp. 687-689, 2009/11/01 2009.

© Springer Fachmedien Wiesbaden GmbH, part of Springer Nature 2019
G. Brankatschk, *Modeling Crop Rotations and Co-Products in Agricultural Life Cycle Assessments*, https://doi.org/10.1007/978-3-658-23588-8

[8]   T. Nemecek, N. Jungbluth, L. M. i Canals, and R. Schenck, "Environmental impacts of food consumption and nutrition: where are we and what is next?," The International Journal of Life Cycle Assessment, vol. 21, no. 5, pp. 607-620, 2016// 2016.

[9]   I. de Boer, "Welcome Address Speech about previous LCA Food conferences: 1996-2016," ed. Dublin, Ireland, 2016.

[10]   B. Notarnicola, S. Sala, A. Anton, S. J. McLaren, E. Saouter, and U. Sonesson, "The role of life cycle assessment in supporting sustainable agri-food systems: A review of the challenges," Journal of Cleaner Production, vol. 140, Part 2, pp. 399-409, 1/1/ 2017.

[11]   S. Lundie, A. Ciroth, and G. Huppes, "Inventory methods in LCA: towards consistency and improvement - Final Report," in "UNEP-SETAC Life Cycle Initiative," Life Cycle Inventory (LCI) Programme Task Force 3: Methodological Consistency, Report June 2007 2007, Available: http://www.estis.net/includes/file.asp?site =lcinit&file=1DBE10DB-888A-4891-9C52-102966464F8D, Accessed on: 26.05.2010.

[12]   N. Pelletier and P. Tyedmers, "An ecological economic critique of the use of market information in life cycle assessment research," Journal of Industrial Ecology, vol. 15, no. 3, pp. 342-354, 2011.

[13]   J. Wrightson, Agriculture theoretical and practical. London: Lockwood, 1921.

[14]   FAO, "SAFA Indicators Sustainability Assessment of Food and Agriculture Systems," ed. Rome, Italy: Food and Agriculture Organization of the United Nations (FAO), 2013.

[15]   G. Brankatschk and M. Finkbeiner, "Application of the Cereal Unit in a new allocation procedure for agricultural life cycle assessments," Journal of Cleaner Production, vol. 73, no. 0, pp. 72-79, 2014.

[16]   G. Brankatschk and M. Finkbeiner, "Modeling crop rotation in agricultural LCAs — Challenges and potential solutions," Agricultural Systems, vol. 138, no. 0, pp. 66-76, 9// 2015.

[17]   G. Brankatschk and M. Finkbeiner, "Crop rotations and crop residues are relevant parameters for agricultural carbon footprints,"

Agronomy for Sustainable Development, vol. 37, no. 6, p. 58, 2017/10/30 2017.

[18]   J. S. Cooper and J. A. Fava, "Life-Cycle Assessment Practitioner Survey: Summary of Results," Journal of Industrial Ecology, vol. 10, no. 4, pp. 12-14, 2006.

[19]   R. Teixeira and S. Pax, "A Survey of Life Cycle Assessment Practitioners with a Focus on the Agri-Food Sector," Journal of Industrial Ecology, vol. 15, no. 6, pp. 817-820, 2011.

[20]   M. Finkbeiner, "Carbon footprinting-opportunities and threats," (in English), International Journal of Life Cycle Assessment, Editorial Material vol. 14, no. 2, pp. 91-94, Mar 2009.

[21]   ISO 14044, "ISO 14044 Environmental Management – Life Cycle Assessment – Requirements and Guidelines," ed. Geneva: International Organization for Standardization (ISO), 2006.

[22]   ISO 14040, "ISO 14040 Environmental Management – Life Cycle Assessment – Principles and Framework," ed. Geneva: International Organization for Standardization (ISO), 2006.

[23]   T. Ekvall and G. Finnveden, "Allocation in ISO 14041--a critical review," Journal of Cleaner Production, vol. 9, no. 3, pp. 197-208, 2001.

[24]   Department for Environment, Food and Rural Affairs, & British Standards Institution, 2011. PAS 2050: 2011-Specification for the Assessment of the Life Cycle Greenhouse Gas Emissions of Goods and Services, 2011.

[25]   (2011). Reading guide for the methodology annex of BP X30-323-0. Available: http://www2.ademe.fr/servlet/getDoc?sort=-1&cid=96& m=3&id=84715&ref=&nocache=yes&p1=111Available: http:// www2.ademe.fr/servlet/getBin?name=8568DE32B5F0A1987A8608B 243207DB9_tomcatlocal1347005677859.pdf

[26]   T. Ekvall and B. Weidema, "System boundaries and input data in consequential life cycle inventory analysis," The International Journal of Life Cycle Assessment, vol. 9, no. 3, pp. 161-171, 2004.

[27]   B. Weidema, "Avoiding Co-Product Allocation in Life-Cycle Assessment," Journal of Industrial Ecology, vol. 4, no. 3, pp. 11-33, 2000.

[28]   B. Weidema, "Market information in life cycle assessment," in "Environmental Project No. 863," Danish Environmental Protection Agency, Copenhagen2003 2003, Accessed on: 23.11.2011.

[29]   P. Pawelzik et al., "Critical aspects in the life cycle assessment (LCA) of bio-based materials – Reviewing methodologies and deriving recommendations," Resources, Conservation and Recycling, vol. 73, no. 0, pp. 211-228, 2013.

[30]   G. Finnveden et al., "Recent developments in Life Cycle Assessment," (in English), Journal of Environmental Management, Review vol. 91, no. 1, pp. 1-21, Oct 2009.

[31]   E. Audsley et al., "Harmonisation of Environmental Life Cycle Assessment for Agriculture: Final report," in "Concerted Action AIR3-CT94-2028," European Commission DG VI Agriculture, Brussels20 August 2003 2003, Accessed on: 26.05.2010.

[32]   A. Mendoza, T. v. Ruijven, K. Vad, and T. Wardenaar, "The Allocation Problem in Bio-Electricity Chains," MSc. Industrial Ecology Master Thesis, Faculty of Science - Institut of Environmental Sciences, Universiteit Leiden, Leiden, 2008.

[33]   M. A. Curran, "Development of life cycle assessment methodology: a focus on co-product allocation," PhD PhD, Erasmus University Rotterdam, 2008.

[34]   S. Kim and B. E. Dale, "Allocation procedure in ethanol production system from corn grain - I. System expansion," (in English), International Journal of Life Cycle Assessment, Article vol. 7, no. 4, pp. 237-243, 2002.

[35]   E. Gnansounou, A. Dauriat, J. Villegas, and L. Panichelli, "Life cycle assessment of biofuels: Energy and greenhouse gas balances," Bioresource Technology, doi: DOI: 10.1016/j.biortech.2009.05.067 vol. 100, no. 21, pp. 4919-4930, 2009.

[36]   A. Singh, D. Pant, N. E. Korres, A.-S. Nizami, S. Prasad, and J. D. Murphy, "Key issues in life cycle assessment of ethanol production from lignocellulosic biomass: Challenges and perspectives," Bioresource Technology, vol. 101, no. 13, pp. 5003-5012, 2010.

[37]   W. Klöpffer and B. Grahl, Ökobilanz (LCA): ein Leitfaden für Ausbildung und Beruf [Life Cycle Assessment (LCA): a guide for training and professional]. Weinheim Wiley-VCH-Verl., 2009, p. 426 Seiten.

[38]   D. Chiaramonti and L. Recchia, "Is life cycle assessment (LCA) a suitable method for quantitative CO_2 saving estimations? the impact of field input on the LCA results for a pure vegetable oil chain," Biomass and Bioenergy, vol. 34, no. 5, pp. 787-797, 2010.

[39]   O. Cavalett and E. Ortega, "Integrated environmental assessment of biodiesel production from soybean in Brazil," (in English), Journal of Cleaner Production, doi: DOI: 10.1016/j.jclepro.2009.09.008 vol. 18, no. 1, pp. 55-70, 2010.

[40]   L. Luo, E. van der Voet, G. Huppes, and H. A. U. de Haes, "Allocation issues in LCA methodology: a case study of corn stover-based fuel ethanol," (in English), International Journal of Life Cycle Assessment, Article vol. 14, no. 6, pp. 529-539, Sep 2009.

[41]   (2012). Erneuerbare Energien in Zahlen - Nationale und internationale Entwicklung [Renewable energy sources in figures - national and international development.]. Available: http://www.bmu.de/service/publikationen/downloads/details/artikel/erneuerbare-energien-in-zahlen/

[42]   OVID. (2013, 27th February 2013). Web page - Section products; http://www.ovid-verband.de/unsere-branche/produkte/ assessed 27th February 2013. Available: http://www.ovid-verband.de/unsere-branche/produkte/

[43]   VDB. (2013, 27th February 2013). Factsheet Biodiesel. Available: www.biokraftstoffverband.de   Available: http://www.biokraftstoff verband.de/index.php/biodiesel.html?file=tl_files/download/Daten_ und_Fakten/factsheet_biodiesel.pdf

[44]   IDF, "A common carbon footprint approach for dairy - The IDF guide to standard lifecycle assessment methodology for the dairy sector," in "Bulletin of the International Dairy Federation," November 2010 2010, Available: http://www.idf-lca-guide.org, Accessed on: 20. März 2013.

[45]   DIRECTIVE 2009/28/EC on the promotion of the use of energy from renewable sources (RED), C. European Parliament Directive, 2009.

[46]   J. W. Becker, "Aggregation in landwirtschaftlichen Gesamtrechnungen über physische Maßstäbe - Futtergersteneinheiten als Generalnenner - [Aggregation in agricultural accounts using physical standards - feed barley units as a common denominator -]," Dissertation Ph.D., Fachbereich Agrarwissenschaften, Justus-Liebig-Universität Gießen, Gießen, 1988.

[47]   S. S. Mönking, C. Klapp, H. Abel, and L. Theuvsen, "Überarbeitung des Getreide- und Vieheinheitenschlüssels - Endbericht zum BMELV-Forschungsprojekt 06HS030 [Revision of cereal unit and livestock unit - Final report on research project 06HS030 BMELV]," Georg-August-Universität Göttingen Fakultät für Agrarwissenschaften Forschungs- und Studienzentrum für Landwirtschaft und Umwelt, Göttingen, Final Report 06HS030, September 2010 2010, Available: http://download.ble.de/06HS030.pdf   Available: http://service.ble.de/fpd_ble/index2.php?detail_id=706&site_key=154&stichw_suche=D UMMY&zeilenzahl_zaehler=1353&NextRow=630 Available: http://service.ble.de/fpd_ble/index2.php?site_key=154&stichw_suche=DU MMY&NextRow=630Available: (Strg+F   "Vieheinheiten...") Available: oder Available: http://www.ble.de/cln_090/nn_467852/DE/04___Forschungsfoerderung/Forschungsfoerderung___node.html? ___nnn=trueAvailable: (Suchbegriff: "Getreideeinheiten"), Accessed on: 29.03.2011.

[48]   C. Klapp, "Getreide- und Vieheinheitenschlüssel als Bewertungsmaßstäbe in der Landwirtschaft: internationaler Vergleich und Konsequenzen alternativer Viehbewertungen [Grain and livestock units as assessment measures in agriculture: international comparison

and implications of alternative livestock reviews] " Ph.D., Cuvillier, Göttingen, 2011.

[49] BMELV, Statistisches Jahrbuch über Ernährung, Landwirtschaft und Forsten der Bundesrepublik Deutschland 2011 [Statistical Yearbook on Food, Agriculture and Forestry of the Federal Republic of Germany 2011]. Landwirtschaftsverlag Münster Hiltrup, 2012.

[50] BMELV. (2013). Getreideeinheitenschlüssel 2012 [List of Cereal Unit conversion factors 2012] (2012 ed.). Available: http://etracker.zadi.de/lnkcnt.php?et=W5E&url=http://berichte.bmelv-statistik.de/SJT-3120100-2012.pdf&lnkname=http://berichte.bmelv-statistik.de/SJT-3120100-2012.pdf

[51] (2008). Leitlinie zur effizienten und umweltverträglichen Erzeugung von Winterweizen [Guidelines for efficient and environmentally sustainable production of winter wheat]. Available: www.tll.de/ainfoAvailable: http://www.thueringen.de/de/tll/pflanzenproduktion/marktfrucht/

[52] (2008). Leitlinie zur effizienten und umweltverträglichen Erzeugung von Winterraps [Guidelines for efficient and environmentally sustainable production of rape seeds]. Available: www.tll.de/ainfo Available: http://www.thueringen.de/de/tll/pflanzenproduktion/marktfrucht/

[53] (2008). Leitlinie zur effizienten und umweltverträglichen Erzeugung von Sommergerste [Guidelines for efficient and environmentally sustainable production of sugar beets]. Available: www.tll.de/ainfo Available: http://www.thueringen.de/de/tll/pflanzenproduktion/marktfrucht/

[54] AGRIDEA and SBV. (2013, 18th February 2013). Agrigate web page http://www.agrigate.ch/de/pflanzenbau/ackerbau/rueben/1316/6702/ accessed 18 February 2013. Available: http://www.agrigate.ch/de/pflanzenbau/660/6702/zuckerruebenblaetter-und-zuckerruebenblattsilage/ Available: http://www.agrigate.ch/de/pflanzenbau/ackerbau/rueben/1316/6702/

[55] M. Kaltschmitt, H. Hartmann, and H. H. (Hrsg.), Energie aus Biomasse: Grundlagen, Techniken und Verfahren [Energy from Biomass: Fundamentals, Techniques and Procedures], 2. ed. Dordrecht; Heidelberg; London; New York: Springer, 2009.

[56] Oil World. (2013) Oil World Monthly February 2013. Oil World monthly. Available: http://www.oilworld.biz

[57] BioGrace. Complete list of standard values, version 3 - Public [Online].   Available: http://www.biograce.net/content/ghgcalculationtools/standardvalues

[58] F. N. R. e. V. FNR, "Basisdaten Bioenergie Deutschland," ed. Gülzow: Fachagentur Nachwachsende Rohstoffe, 2013, p. 48.

[59] FAO, The state of food and agriculture 2009 : livestock in the balance (The State of Food and Agriculture). Rome: Food and Agriculture Organization of the United Nations (FAO), 2009.

[60] C. Cederberg and B. Mattsson, "Life cycle assessment of milk production -- a comparison of conventional and organic farming," (in english), Journal of Cleaner Production, vol. 8, no. 1, pp. 49-60, 2000.

[61] C. Cederberg and M. Stadig, "System Expansion and Allocation in Life Cycle Assessment of Milk and Beef Production," (in English), International Journal of Life Cycle Assessment, Article vol. 8, no. 6, pp. 350 - 356 July 2003.

[62] T. T. H. Nguyen et al., "Effect of dairy production system, breed and co-product handling methods on environmental impacts at farm level," Journal of Environmental Management, vol. 120, no. 0, pp. 127-137, 2013.

[63] A. J. Feitz, S. Lundie, G. Dennien, M. Morain, and M. Jones, "Generation of an industry-specific physico-chemical allocation matrix - Application in the dairy industry and implications for systems analysis," (in English), International Journal of Life Cycle Assessment, Article vol. 12, no. 2, pp. 109-117, Mar 2007.

[64] N. Sarkar, S. K. Ghosh, S. Bannerjee, and K. Aikat, "Bioethanol production from agricultural wastes: An overview," (in English),

Renewable Energy, doi: 10.1016/j.renene.2011.06.045 vol. 37, no. 1, pp. 19-27, 2012.

[65]   Y. Sun and J. Cheng, "Hydrolysis of lignocellulosic materials for ethanol production: a review," (in English), Bioresource Technology, doi: 10.1016/S0960-8524(01)00212-7 vol. 83, no. 1, pp. 1-11, 2002.

[66]   F. Talebnia, D. Karakashev, and I. Angelidaki, "Production of bioethanol from wheat straw: An overview on pretreatment, hydrolysis and fermentation," (in English), Bioresource Technology, doi: 10.1016/j.biortech.2009.11.080 vol. 101, no. 13, pp. 4744-4753, 2010.

[67]   M. Wang, J. Wang, and J. X. Tan, "Lignocellulosic Bioethanol: Status and Prospects," (in english), Energy Sources, Part A: Recovery, Utilization, and Environmental Effects, doi: 10.1080/15567030903226249 vol. 33, no. 7, pp. 612-619, 2011/01/31 2010.

[68]   FNR. (2012, June 2012). 2nd International Symposium on Straw Energy 29 and 30 March 2012. Available: http://www.fnr-server.de/cms35/index.php?id=5687

[69]   I. Esenwein-Rothe, "Zur Methodik der statistischen Aggregation [The methodology of the statistical aggregation]," (in German), Statistische Hefte, vol. 8, no. 1, pp. 66-78, 1967/03/01 1967.

[70]   M. Besch and E. Wöhlken, Zielsetzung, Aussagemöglichkeiten und Aussagegrenzen von mengen-und wertmäßigen Gesamtrechnungen [Purpose, Meaningfulness and Limitations of Quantitative and Value Total Accounts]. Statistisches Amt der Europäischen Gemeinschaften, 1973.

[71]   H. Schneeweiß, "Das Aggregationsproblem [The problem of aggregation]," (in German), Statistische Hefte, vol. 6, no. 1, pp. 1-26, 1965/12/01 1965.

[72]   BMELV. (2013). Getreideeinheitenschlüssel 2012 (2012 ed.). Available: http://etracker.zadi.de/lnkcnt.php?et=W5E&url=http:// berichte.bmelv-statistik.de/SJT-3120100-2012.pdf&lnkname=http:// berichte.bmelv-statistik.de/SJT-3120100-2012.pdf

[73]   M. Kirchgeßner, Tierernährung: Leitfaden für Studium, Beratung und
        Praxis [Animal Nutrition: Guidelines for academic studies, consulting
        and practical application]. Frankfurt am Main: DLG-Verl, 2008.

[74]   R. Jarrige, Ruminant nutrition : recommended allowances and feed
        tables. London [u.a.]: Libbey [u.a.], 1989.

[75]   K. Hayashi, G. Gaillard, and T. Nemecek, "Life cycle assessment of
        agricultural   production   systems: current   issues   and   future
        perspectives," Good agricultural practice (GAP) in Asia and Oceania.
        Food and Fertilizer Technology Center, Taipei, pp. 98-110, 2006.

[76]   A. Mourad, L. Coltro, P. P. L. V. Oliveira, R. Kletecke, and J. A.
        Baddini, "A simple methodology for elaborating the life cycle
        inventory of agricultural products," (in English), The International
        Journal of Life Cycle Assessment, vol. 12, no. 6, pp. 408-413,
        2007/09/01 2007.

[77]   M. Núñez Pineda, "Modelling Location-dependent Environmental
        Impacts in Life Cycle Assessment: Water Use, Desertification and Soil
        Erosion: Application to Energy Crops Grown in Spain," 2011.

[78]   P. Roy et al., "A review of life cycle assessment (LCA) on some food
        products," Journal of Food Engineering, vol. 90, no. 1, pp. 1-10, 2009.

[79]   E. M. Schau and A. M. Fet, "LCA studies of food products as
        background for environmental product declarations," (in English),
        International Journal of Life Cycle Assessment, Review vol. 13, no. 3,
        pp. 255-264, May 2008.

[80]   F. Cherubini and A. H. Strømman, "Life cycle assessment of bioenergy
        systems: State of the art and future challenges," Bioresource
        Technology, vol. 102, no. 2, pp. 437-451, 2011.

[81]   H. M. G. van der Werf, T. Garnett, M. S. Corson, K. Hayashi, D.
        Huisingh, and C. Cederberg, "Towards eco-efficient agriculture and
        food systems: theory, praxis and future challenges," Journal of
        Cleaner Production, vol. 73, no. 0, pp. 1-9, 2014.

[82]   S. Cowell, R. R. Clift, and F. Society, "Life cycle assessment for food
        production systems / by Sarah J. Cowell and Roland Clift," 1995:
        Fertiliser Society.

[83] B. C. Ball, I. Bingham, R. M. Rees, C. A. Watson, and A. Litterick, "The role of crop rotations in determining soil structure and crop growth conditions," Canadian Journal of Soil Science, vol. 85, no. 5, pp. 557-577, 2005/11/01 2005.

[84] D. L. Karlen, C. A. Cambardella, J. L. Kovar, and T. S. Colvin, "Soil quality response to long-term tillage and crop rotation practices," Soil and Tillage Research, vol. 133, no. 0, pp. 54-64, 2013.

[85] E. Curl, "Control of plant diseases by crop rotation," (in English), The Botanical Review, vol. 29, no. 4, pp. 413-479, 1963/10/01 1963.

[86] R. Forsyth, The principles and practice of agriculture, systematically explained: in two volumes: being a treatise compiled for the fourth edition of the Encyclopaedia Britannica. Printed by the proprietor, A Bell, 1804.

[87] C. Daubeny, "Memoir on the Rotation of Crops, and on the Quantity of Inorganic Matters Abstracted from the Soil by Various Plants under Different Circumstances," Philosophical Transactions of the Royal Society of London, vol. 135, no. ArticleType: research-article / Full publication date: 1845 /, pp. 179-252, 1845.

[88] J. Liebig, L. P. Playfair, and J. W. Webster, Chemistry in its application to agriculture and physiology, Third Edition ed. J. Owen, 1842.

[89] W. Abel, Geschichte der deutschen Landwirtschaft vom frühen Mittelalter bis zum 19. Jahrhundert. Stuttgart: Ulmer, 1978.

[90] J. Wirghtson, Agriculture theoretical and practical. London: Lockwood, 1921.

[91] D. G. Bullock, "Crop rotation," Critical Reviews in Plant Sciences, vol. 11, no. 4, pp. 309-326, 1992/01/01 1992.

[92] Z. Berzsenyi, B. Győrffy, and D. Lap, "Effect of crop rotation and fertilisation on maize and wheat yields and yield stability in a long-term experiment," European Journal of Agronomy, vol. 13, no. 2–3, pp. 225-244, 2000.

[93] E. J. Russell and E. W. Russell, Soil conditions and plant growth. Longman, 1973.

[94]   H. Blanco-Canqui and R. Lal, "Crop Residue Removal Impacts on Soil Productivity and Environmental Quality," Critical Reviews in Plant Sciences, doi: 10.1080/07352680902776507 vol. 28, no. 3, pp. 139-163, 2009/04/03 2009.

[95]   L. J. Munkholm, R. J. Heck, and B. Deen, "Long-term rotation and tillage effects on soil structure and crop yield," Soil and Tillage Research, vol. 127, no. 0, pp. 85-91, 2013.

[96]   D. L. Karlen, G. E. Varvel, D. G. Bullock, and R. M. Cruse, "Crop Rotations for the 21st Century," in Advances in Agronomy, vol. Volume 53, L. S. Donald, Ed.: Academic Press, 1994, pp. 1-45.

[97]   W. Zegada-Lizarazu and A. Monti, "Energy crops in rotation. A review," Biomass and Bioenergy, vol. 35, no. 1, pp. 12-25, 2011.

[98]   T. Nemecek and J. Schnetzer, "Data collection of inputs and yields in LCIs of agricultural production systems in Switzerland and other European countries - Data v3.0 (2012)," Agroscope Reckenholz-Tänikon Research Station ART, ZurichAugust 2011 2011, Available: https://ecoquery.ecoinvent.org/File/ReportsAvailable: https:// ecoquery.ecoinvent.org/File/File?fileName=01%20crop%20production %20-%20yields%20inputs%20CH%20Europe%20v1.1.pdf&hash=-2031344868&type=Reports, Accessed on: 14. August 2013.

[99]   T. Nemecek, D. Dubois, O. Huguenin-Elie, and G. Gaillard, "Life cycle assessment of Swiss farming systems: I. Integrated and organic farming," Agricultural Systems, vol. 104, no. 3, pp. 217-232, 2011.

[100] T. Nemecek and T. Kägi, "Life Cycle Inventories of Agricultural Production Systems; ecoinvent report No. 15," in "ecoinvent report," ecoinvent Swiss Centre for Life Cycle Inventories, Zürich and DübendorfDecember 2007 2007, Available: www.ecoinvent.org Available: www.art.admin.ch, Accessed on: 17. August 2013.

[101] T. Nemecek, J.-S. von Richthofen, G. Dubois, P. Casta, R. Charles, and H. Pahl, "Environmental impacts of introducing grain legumes into European crop rotations," European Journal of Agronomy, vol. 28, no. 3, pp. 380-393, 2008.

[102] H. van Zeijts, H. Leneman, and A. Wegener Sleeswijk, "Fitting fertilisation in LCA: allocation to crops in a cropping plan," Journal of Cleaner Production, vol. 7, no. 1, pp. 69-74, 1999.

[103] J. Martínez-Blanco et al., "Compost benefits for agriculture evaluated by life cycle assessment. A review," (in English), Agronomy for Sustainable Development, vol. 33, no. 4, pp. 721-732, 2013/10/01 2013.

[104] J. Martínez-Blanco et al., "Assessing the Environmental Benefits of Compost Use-on-Land through an LCA Perspective," in Sustainable Agriculture Reviews, vol. 12, E. Lichtfouse, Ed. (Sustainable Agriculture Reviews: Springer Netherlands, 2013, pp. 255-318.

[105] J. Martínez-Blanco, J. Rieradevall, A. Antón, and P. Muñoz, "Multifunctionality-solving approaches of compost application in crop rotations," Journal of Cleaner Production, vol. 64, no. 0, pp. 384-395, 2014.

[106] E. Audsley, S. Alber, and E. Gemeinschaften, Harmonisation of environmental life cycle assessment for agriculture. European Comm., DG VI Agriculture, 1997.

[107] T. Alföldi, O. Schmid, G. Gaillard, and D. Dubois, "IP-und Bio-Produktion: Ökobilanzierung über eine Fruchtfolge," Agrarforschung, vol. 6, no. 9, pp. 337-340, 1999.

[108] R. Flisch, S. Sinaj, R. Charles, and W. Richner, "GRUDAF 2009. Principles for fertilisation in arable and fodder production," Agrarforschung, vol. 16, no. 2, pp. 1-100, 2009.

[109] Proposal for a Directive Of The European Parliament And Of The Council amending Directive 98/70/EC relating to the quality of petrol and diesel fuels and amending Directive 2009/28/EC on the promotion of the use of energy from renewable sources, European Commission, 2012.

[110] M. S. Castellazzi, G. A. Wood, P. J. Burgess, J. Morris, K. F. Conrad, and J. N. Perry, "A systematic representation of crop rotations," Agricultural Systems, vol. 97, no. 1-2, pp. 26-33, 2008.

[111] D. R. Cox and H. D. Miller, The Theory of Stochastic Processes, 1st
       ed. London and Colchester: Birbeck College, University of London,
       1965.

[112] K. f. T. u. B. i. d. L. e. V. KTBL. Verfahrensrechner Pflanze
       [Online]. Available: http://daten.ktbl.de/vrpflanze/prodverfahren

[113] K. f. r. T. u. B. i. d. L. e. V. KTBL, Betriebsplanung Landwirtschaft
       2010/11 : Daten für die Betriebsplanung in der Landwirtschaft, 22 ed.
       Darmstadt: Kuratorium für Technik und Bauwesen in der
       Landwirtschaft, 2010.

[114] B. f. E. u. L. BMEL and B. f. L. u. E. BLE, "Besondere Ernte- und
       Qualitätsermittlung (BEE) 2013," in "Reihe: Daten-Analysen,"
       Bundesministerium für Ernährung und Landwirtschaft BMEL,
       BerlinApril 2014 2014, Available: http://www.bmelv-statistik.de/
       de/fachstatistiken/besondere-ernteermittlung/Available: http://
       berichte.bmelv-statistik.de/EQB-1002000-2013.pdf, Accessed on: 15.
       Februar 2015.

[115] BioGrace. BioGrace Excel tool - version 4c [Online]. Available:
       http://www.biograce.net/img/files/2013-10-09-123912BioGrace_
       GHG_calculation_tool_-_version_4c.zip   Available: http://
       www.biograce.net/home

[116] G. Baumgärtel, M. Benke, and T. Eiler, "Düngeempfehlungen
       Stickstoff   Getreide   Raps   Hackfrüchte   Mais,"   ed:
       Landwirtschaftskammer Niedersachsen, 2010, p. 2.

[117] FAO, "Breakthrough climate agreement recognizes food security as a
       priority," C. Emdsen, Ed., ed. Rome, Italy: Food and Agriculture
       Organization of the United Nations FAO, 2015, p. 2.

[118] FAO, ""Climate-Smart" Agriculture – Policies, Practices and
       Financing for Food Security, Adaptation and Mitigation," FAO Food
       and Agriculture Organization of the United Nations, Rome2010,
       Available: http://www.fao.org/docrep/013/i1881e/i1881e00.htm
       Available: http://www.fao.org/docrep/013/i1881e/i1881e00.pdf,
       Accessed on: 25 December 2015.

[119] FAO, "Climate-Smart Agriculture – Sourcebook," FAO Food and Agriculture Organization of the United Nations, Rome2013, Available: http://www.fao.org/climate-smart-agriculture/72611/en/ Available: http://www.fao.org/docrep/018/i3325e/i3325e.pdf, Accessed on: 25 December 2015.

[120] L. Lipper et al., "Climate-smart agriculture for food security," Nature Clim. Change, Perspective vol. 4, no. 12, pp. 1068-1072, 12//print 2014.

[121] United Nations, "World Population Prospects: The 2015 Revision," United Nations, Department of Economic and Social Addairs, Population Division2015, Available: http://esa.un.org/unpd/wpp/, Accessed on: 05. September 2015.

[122] FAO. FAO Statistics Division - Arable Land World [Online]. Available: http://faostat3.fao.org/download/R/RL/E

[123] J. H. J. Spiertz, "Nitrogen, sustainable agriculture and food security. A review," Agronomy for Sustainable Development, vol. 30, no. 1, pp. 43-55, 2010// 2010.

[124] C. Liu, H. Cutforth, Q. Chai, and Y. Gan, "Farming tactics to reduce the carbon footprint of crop cultivation in semiarid areas. A review," Agronomy for Sustainable Development, vol. 36, no. 4, p. 69, 2016// 2016.

[125] A. Wezel, G. Soboksa, S. McClelland, F. Delespesse, and A. Boissau, "The blurred boundaries of ecological, sustainable, and agroecological intensification: a review," Agronomy for Sustainable Development, vol. 35, no. 4, p. 0, 2015// 2015.

[126] C. Bessou, F. Ferchaud, B. Gabrielle, and B. Mary, "Biofuels, greenhouse gases and climate change. A review," Agronomy for Sustainable Development, vol. 31, no. 1, p. 1, 2011// 2011.

[127] J. Dury, N. Schaller, F. Garcia, A. Reynaud, and J. E. Bergez, "Models to support cropping plan and crop rotation decisions. A review," Agronomy for Sustainable Development, vol. 32, no. 2, pp. 567-580, 2012// 2012.

[128] S. Cowell and R. Clift, "Life cycle assessment for food production systems," ed: Fertiliser Society, 1995.

[129] M. A. Curran, "Studying the Effect on System Preference by Varying Coproduct Allocation in Creating Life-Cycle Inventory," (in English), Environmental Science & Technology, vol. 41, no. 20, pp. 7145-7151, 09/18 2007.

[130] A. Berthoud, A. l. Buet, T. Genter, and S. Marquis, "Comparison of the environmental impact of three forms of nitrogen fertilizer," Fertilizers Europe, Paris2012, Available: http:// fertilizerseurope.com/fileadmin/user_upload/publications/agriculture _publications/Enviro_Impact-V9.pdf, Accessed on: 06. Februar 2016.

[131] J. Braschkat, A. Patyk, M. Quirin, and G. Reinhardt, "Life cycle assessment of bread production-a comparison of eight different scenarios," in Life Cycle Assessment in the Agri-food sector. Proceedings from the "4th International Conference, Bygholm (DK), 6 - 8 October 2003, Bygholm, Denmark, 2004, vol. 9.

[132] M. Müller-Lindenlauf, C. Cornelius, S. Gärtner, G. Reinhardt, N. Rettenmaier, and T. Schmidt, "Umweltbilanz von Milch- und Milcherzeugnissen - Status quo und Ableitung von Optimierungspotenzialen," ifeu – Institut für Energie- und Umweltforschung Heidelberg GmbH, Heidelberg31.Oktober 2014 2014, Available: http://www.milchindustrie.de/aktuelles/ pressemitteilungen/umweltbilanz-von-milch-und-milcherzeugnissen-erstellt/Available: http://www.milchindustrie.de/uploads/tx_news/ IFEU-VDM-Milchbericht-Umweltbilanz-2014_01.pdf, Accessed on: 06. Februar 2016.

[133] BioGrace. BioGrace Excel tool - version 4d - Harmonised Calculations of Biofuel Greenhouse Gas Emissions in Europe [Online]. Available: http://www.biograce.net Available: http:// www.biograce.net/content/ghgcalculationtools/recognisedtool/ Available: http://www.biograce.net/img/files/2015-05-12-161933Bio Grace-I_GHG_calculation_tool_-_version_4d.zip

[134] P. Goglio, G. Brankatschk, M. T. Knudsen, A. G. Williams, and T. Nemecek, "Addressing crop interactions within cropping systems in LCA," The International Journal of Life Cycle Assessment, 2017/09/08 2017.

[135] R. Łukowiak, W. Grzebisz, and G. F. Sassenrath, "New insights into phosphorus management in agriculture — A crop rotation approach," Science of The Total Environment, vol. 542, Part B, pp. 1062-1077, 1/15/ 2016.

[136] S. Kim and B. E. Dale, "Global potential bioethanol production from wasted crops and crop residues," Biomass and Bioenergy, vol. 26, no. 4, pp. 361-375, 4// 2004.

[137] R. Lal, "Soil quality impacts of residue removal for bioethanol production," Soil and Tillage Research, vol. 102, no. 2, pp. 233-241, 3// 2009.

[138] R. Lal, "World crop residues production and implications of its use as a biofuel," Environment International, vol. 31, no. 4, pp. 575-584, 5// 2005.

[139] B. Gabrielle and N. Gagnaire, "Life-cycle assessment of straw use in bio-ethanol production: A case study based on biophysical modelling," Biomass and Bioenergy, vol. 32, no. 5, pp. 431-441, 5// 2008.

[140] H. C. von Carlowitz, Sylvicultura oeconomica. Leipzig: Johann Friedrich Braun, 1713, p. 460.

[141] P. Koch and T. Salou, "AGRIBALYSE®: Rapport Méthodologique – Version 1.3," ADEME, Angers, FranceNovember 2016 2016, Available: http://www.ademe.fr/en/expertise/alternative-approaches-to-production/agribalyse-programAvailable: http://www.ademe.fr/sites/default/files/assets/documents/agribalyse_v1_3_methodology. pdf Available: http://www.ademe.fr/sites/default/files/assets/documents/agribalyse_fs_v1_3.xlsx Available: http://www.ademe.fr/sites/default/files/assets/documents/agribalyse_v1_3_report_of_c hanges_2016.pdf, Accessed on: 24. Juni 2017.

[142] T. Nemecek, O. Huguenin-Elie, D. Dubois, G. Gaillard, B. Schaller, and A. Chervet, "Life cycle assessment of Swiss farming systems: II.

Extensive and intensive production," Agricultural Systems, vol. 104, no. 3, pp. 233-245, 2011.

[143] T. Alföldi, O. Schmid, G. Gaillard, and D. Dubois, "Life cycle assessment of integrated and organic crop production," Agrarforschung, vol. 6, no. 9, pp. 337-340, 1999.

[144] N. Pelletier, "Life Cycle Thinking, Measurement and Management for Food System Sustainability," Environmental Science & Technology, vol. 49, no. 13, pp. 7515-7519, 2015/07/07 2015.

[145] M. Finkbeiner, E. M. Schau, A. Lehmann, and M. Traverso, "Towards Life Cycle Sustainability Assessment," Sustainability, vol. 2, no. 10, p. 3309, 2010.

[146] J. B. Guinée et al., "Life Cycle Assessment: Past, Present, and Future," Environmental Science & Technology, vol. 45, no. 1, pp. 90-96, 2011/01/01 2011.

[147] M. Finkbeiner et al., "Challenges in Life Cycle Assessment: An Overview of Current Gaps and Research Needs," in Background and Future Prospects in Life Cycle Assessment, W. Klöpffer, Ed. (LCA Compendium – The Complete World of Life Cycle Assessment: Springer Netherlands, 2014, pp. 207-258.

[148] BMEL and BLE, "Besondere Ernte- und Qualitätsermittlung BEE 2014 (Special harvesting and quality determination BEE 2014)," in "Reihe: Daten-Analysen," BMEL Bundesministerium für Ernährung und Landwirtschaft (Federal Ministry of Food and Agriculture) BLE Bundesanstalt für Landwirtschaft und Ernährung (Federal Office for Food and Agriculture), Berlin2015, Available: http://www.bmelv-statistik.de/de/fachstatistiken/besondere-ernteermittlung/ Available: http://berichte.bmelv-statistik.de/EQB-1002000-2014.pdf, Accessed on: 04. Mai 2015.

[149] Eurostat. Crops products - annual data [Online]. Available: http://appsso.eurostat.ec.europa.eu/

[150] LfL, "Basisdaten fuer die Ermittlung des Duengebedarfs, fuer die Umsetzung der Duengeverordung,zur Berechnung des KULAP-Naaehrstoff-Saldos, zur Berechnung der Nährstoffbilanz nach Hoftor-

Ansatz (Basic data: for the determination of nutrient demand, for the implementation of fertilizer ordinance, to calculate the KULAP-nutrient-balance, to calculate the nutrient balance using farm-gate approach)," ed: LfL Bayerische Landesanstalt für Lanwirtschaft (The Bavarian State Research Center for Agriculture), 2013, p. 25.

[151] KTBL. Verfahrensrechner Pflanze (Calculator for crop production processes) [Online]. Available: http://daten.ktbl.de/vrpflanze/prodverfahren

[152] (2010). Leitlinie zur effizienten und umweltverträglichen Erzeugung von Ackerbohnen und Körnererbsen (Guideline for efficient and environment-friendly production of field beans and peas). Available: http://www.tll.de/ainfo/pdf/ll_abke.pdf

[153] R. Albrecht and C. Guddat, "Welchen Wert haben Körnerleguminosen in der Fruchtfolge (What is the value of grain legumes in crop rotations)," ed, 2004.

[154] Eurostat and European Union, The use of plant protection products in the European Union : data 1992-2003. Luxembourg: EUR-OP, 2007.

[155] J.-S. v. Richthofen et al., "Economic interest of grain legumes in European crop rotations," in "European extension network for the development of grain legumes production in the EU "GL-Pro"," 2006, Accessed on: 10. April 2015.

[156] finanzen.net. (2015, 21 June 2015). finanzen.net: Börse und Finanzen. Available: www.finanzen.net

[157] www.agrarheute.com. (2015, 21 June 2015). Nachrichten für die Landwirtschaft | agrarheute.com. Available: http://www.agrarheute.com/markt-uebersicht Available: http://www.agrarheute.com/raps Available: http://www.agrarheute.com/gerste-551190 Available: http://www.agrarheute.com/erzeugerpreise-stroh

[158] M. Finkbeiner, "Product environmental footprint—breakthrough or breakdown for policy implementation of life cycle assessment?," ed: Springer, 2014.

[159] A. Lehmann, V. Bach, and M. Finkbeiner, "Product environmental footprint in policy and market decisions: applicability and impact assessment," Integrated environmental assessment and management, vol. 11, no. 3, pp. 417-424, 2015.

[160] C. Cederberg and M. Stadig, "System Expansion and Allocation in Life Cycle Assessment of Milk and Beef Production," (in English), International Journal of Life Cycle Assessment, Article vol. 8, no. 6, pp. 350 - 356, July 2003.

[161] B. Notarnicola, G. Tassielli, P. A. Renzulli, and F. Monforti, "Energy flows and greenhouses gases of EU (European Union) national breads using an LCA (Life Cycle Assessment) approach," Journal of Cleaner Production, vol. 140, Part 2, pp. 455-469, 1/1/ 2017.

[162] H. M. G. van der Werf and T. T. H. Nguyen, "Construction cost of plant compounds provides a physical relationship for co-product allocation in life cycle assessment," The International Journal of Life Cycle Assessment, journal article vol. 20, no. 6, pp. 777-784, 2015.

[163] X. Chen, A. Wilfart, L. Puillet, and J. Aubin, "A new method of biophysical allocation in LCA of livestock co-products: modeling metabolic energy requirements of body-tissue growth," The International Journal of Life Cycle Assessment, vol. 22, no. 6, pp. 883-895, 2017// 2017.

[164] N. Palmieri, M. B. Forleo, G. Giannoccaro, and A. Suardi, "Environmental impact of cereal straw management: An on-farm assessment," Journal of Cleaner Production, vol. 142, Part 4, pp. 2950-2964, 1/20/ 2017.

[165] R. Łukowiak, W. Grzebisz, and P. Barłóg, "Magnesium management in the soil-crop system–a crop rotation approach," Plant Soil Environ., vol. 62, no. 9, pp. 395-401, 2016.

[166] M. Macák, Š. Žák, and M. Andrejčíková, "Productivity and macroelements content of cereal and legume crops," Acta Fytotechnica et Zootechnica, vol. 18, no. 5, pp. 160-162, 2016.

[167] N. Palmieri, M. B. Forleo, and E. Salimei, "Environmental impacts of a dairy cheese chain including whey feeding: An Italian case study," Journal of Cleaner Production, vol. 140, Part 2, pp. 881-889, 1/1/ 2017.

[168] C. Tricase, E. Lamonaca, C. Ingrao, J. Bacenetti, and A. Lo Giudice, "A comparative Life Cycle Assessment between organic and conventional barley cultivation for sustainable agriculture pathways," Journal of Cleaner Production, 2017/07/03/ 2017.

[169] T. J. Dijkman, M. Birkved, H. Saxe, H. Wenzel, and M. Z. Hauschild, "Environmental impacts of barley cultivation under current and future climatic conditions," Journal of Cleaner Production, vol. 140, Part 2, pp. 644-653, 1/1/ 2017.

[170] S. G. Mackenzie, I. Leinonen, and I. Kyriazakis, "The need for co-product allocation in the life cycle assessment of agricultural systems—is "biophysical" allocation progress?," The International Journal of Life Cycle Assessment, vol. 22, no. 2, pp. 128-137, 2017// 2017.

[171] C. Peter, K. Helming, and C. Nendel, "Do greenhouse gas emission calculations from energy crop cultivation reflect actual agricultural management practices? – A review of carbon footprint calculators," Renewable and Sustainable Energy Reviews, vol. 67, pp. 461-476, 1// 2017.

[172] M. I. S. F. Matsuura, F. R. T. Dias, J. F. Picoli, K. R. G. Lucas, C. de Castro, and M. H. Hirakuri, "Life-cycle assessment of the soybean-sunflower production system in the Brazilian Cerrado," The International Journal of Life Cycle Assessment, vol. 22, no. 4, pp. 492-501, 2017// 2017.

[173] C. Peter et al., "Impact of Energy Crop Rotation Design on Multiple Aspects of Resource Efficiency," Chemical Engineering & Technology, vol. 40, no. 2, pp. 323-332, 2017.

[174] U. E. Prechsl, R. Wittwer, M. G. A. van der Heijden, G. Lüscher, P. Jeanneret, and T. Nemecek, "Assessing the environmental impacts of cropping systems and cover crops: Life cycle assessment of FAST, a

long-term arable farming field experiment," Agricultural Systems, vol. 157, pp. 39-50, 10// 2017.

[175] C. Peter et al., "The MiLA tool: Modeling greenhouse gas emissions and cumulative energy demand of energy crop cultivation in rotation," Agricultural Systems, vol. 152, pp. 67-79, 3// 2017.

[176] F. Capitanescu, A. Marvuglia, T. Navarrete Gutiérrez, and E. Benetto, "Multi-stage farm management optimization under environmental and crop rotation constraints," Journal of Cleaner Production, vol. 147, pp. 197-205, 3/20/ 2017.

[177] P. Goglio et al., "Development of Crop.LCA, an adaptable screening life cycle assessment tool for agricultural systems: A Canadian scenario assessment," Journal of Cleaner Production, 2017.

[178] UN, "Transforming our world: the 2030 Agenda for Sustainable Development," U. N. G. Assembly, Ed., ed: United Nations (UN), 2015.

[179] UN, "Resolution adopted by the General Assembly: 55/2. United Nations Millennium Declaration," ed: United Nations, 2000, p. 9.

[180] UN, "Resolution adopted by the General Assembly on 16 September 2005," ed: United Nations (UN), 2005, p. 38.

[181] UN, "The Millennium Development Goals Report 2015," ed: United Nations (UN), 2015, p. 75.

[182] FAO, IFAD, and WFP, "The State of Food Insecurity in the World 2015," Food and Agriculture Organization of the United Nations (FAO), International Fund for Agricultural Development (IFAD), World Food Programme (WFP)2015, Accessed on: 10. Juni 2016.

[183] United Nations Framework Convention on Climate Change UNFCCC. (2015). FCCC/CP/2015/L.9/Rev.1, Adoption of the Paris Agreement. Available: http://unfccc.int/documentation/documents/ advanced_search/items/6911.php?priref=600008831Available: http://unfccc.int/resource/docs/2015/cop21/eng/l09r01.pdf

[184] IPPC, "Climate Change 2014: Mitigation of Climate Change - Working Group III Contribution to the Fifth Assessment Report of the Intergovernmental Panel on Climate Change," O. Edenhofer, R.

Pichs-Madruga, Y. Sokona, E. Farahani, S. Kadner, K.Seyboth, A. Adler, I. Baum, S. Brunner, P. Eickemeier, B. Kriemann, J. Savolainen, S. Schlömer, C. von Stechow, T. Zwickel and J.C. Minx (eds.), Ed., ed. Cambridge, United Kindom and New York, USA: Cambridge University Press, 2014, p. 1454.

[185] UNEP, "The Emissions Gap Report," U. N. E. P. (UNEP), Ed., ed. Nairobi, 2015.

[186] UNEP. (2016, 10th July 2016). UNEP-live - Emissions-Impacts-Climate Change. Available: http://uneplive.org/theme/index/13#indcs

[187] UN-REDD. (2016, 10th July 2016). About UN-REDD Programme Collaborative online workspace. Available: http://www.unredd.net/index.php?option=com_content&view=article&id=2082&Itemid=515

[188] World Economic Forum, "The Global Challenge on Food Security and Agriculture - A global initiative of the World Economic Forum," ed, 2016.

[189] World Economic Forum. (2016, 10th July 2016). New Vision for Agriculture. Available: https://www.weforum.org/projects/new-vision-for-agriculture/

[190] FAO. (2016, 10t July 2016). Sustainable Food and Agriculture - Frameworks and approaches. Available: http://www.fao.org/sustainability/frameworks-approaches/en/

[191] M. Rai, T. Reeves, S. Pandey, and L. Collette, "Save and grow: a policymaker's guide to sustainable intensification of smallholder crop production," Rome: FAO, 2011.

[192] FAO, "Second Global Plan of Action for Plant Genetic Resources for Food and Agriculture," ed: Food and Agriculture Organization of the United Nations (FAO), 2016.

[193] FAO. (2016, 10th July 2016). Sustainable Land Management. Available: http://www.fao.org/nr/land/sustainable-land-management/en/

[194] FAO, "Economics and Policy Innovations for Climate-Smart Agriculture," vol. 2016, ed: FAO Food and Agricuture Organization of the United Nations, 2016.

[195] FAO. (2016, 10th July 2016). Integrating Agriculture in National Adaptation Plans Programme (NAPs). Available: http://www.fao.org/climate-change/programmes-and-projects/detail/en/c/328984/

[196] FAO. (2016, 10th July 2016). Livestock Environmental Assessment and Performance (LEAP) Partnership. Available: http://www.fao.org/climate-change/programmes-and-projects/detail/en/c/328987/

[197] FAO. (2016, 10th July 2016). Sustainability Assessment of Food and Agriculture systems (SAFA). Available: http://www.fao.org/climate-change/programmes-and-projects/detail/en/c/328975/

[198] N. Scialabba, SAFA guidelines - sustainability assessment of food and agriculture systems. Food and Agriculture Organization of the United Nations; Rome: FAO, July 2014, 2014.

[199] UN, "Report of the United Nations Conference on the Human Environment - Stockholm 5-16 June 1972," ed: United Nations (UN), 1972, p. 80.

[200] IUCN, UNEP, WWF, FAO, and UNESCO, "World Conservation Strategy - Living Resource Consercation for Sustainable Development," International Union for Conservation of Nature and Natural Resources (IUCN), United Nations Environment Programme (UNEP), World Wildlife Fund (UNEP), Food and Agriculture Organization of the United Nations (FAO), United Nations Educational, Scientific and Cultural Organization (UNESCO), Gland, Switzerland1980, Available: https://portals.iucn.org/library/efiles/html/wcs-004/cover.html Available: https://portals.iucn.org/library/efiles/edocs/WCS-004.pdf.

[201] G. Brundtland et al., "Report of the World Commission on Environment and Development: Our Common Future," (in English), p. 300, 1987.

[202] FAO, The state of food and agriculture. Food and Agriculture Organization of the United Nations (FAO), 1989.

[203] UN, "Report of the United Nations Conference on Environment and Development - Rio de Janeiro, 3-14 June 1992," ed. New York: United Nations (UN), 1993, p. 492.

[204] UNEP/SETAC, "Life Cycle Apporaches The road from analysis to practice," ed: UNEP/SETAC Life Cycle Initiative; United Nations Environment Programme (UNEP); Society of Environmental Toxicology and Chemistry (SETAC), 2005, p. 89.

[205] UNEP/SETAC, "Towards a Life Cycle Sustainability Assessment - Making informed choices on products," ed: UNEP/SETAC Life Cycle Initiative; United Nations Environment Programme (UNEP); Society for Environmental Toxicology and Chemistry (SETAC), 2011, p. 86.

[206] W. Klöpffer, "Life-Cycle based methods for sustainable product development," The International Journal of Life Cycle Assessment, vol. 8, no. 3, pp. 157-159, 2003// 2003.

Annex — Online Tables

The Annex can be accessed at extras.springer.com.

© Springer Fachmedien Wiesbaden GmbH, part of Springer Nature 2019
G. Brankatschk, *Modeling Crop Rotations and Co-Products in Agricultural Life Cycle Assessments*, https://doi.org/10.1007/978-3-658-23588-8

Printed in the United States
By Bookmasters